THE JOHN DEERE CENTURY

RANDY LEFFINGWELL

CRESTLINE

Inspiring | Educating | Creating | Entertaining

Brimming with creative inspiration, how-to projects, and useful information to enrich your everyday life, Quarto Knows is a favorite destination for those pursuing their interests and passions. Visit our site and dig deeper with our books into your area of interest: Quarto Creates, Quarto Cooks, Quarto Homes, Quarto Lives, Quarto Drives, Quarto Explores, Quarto Gifts, or Quarto Kids.

This edition published in 2020 by Crestline,
an imprint of The Quarto Group
142 West 36th Street, 4th Floor
New York, NY 10018 USA
T (212) 779-4972 F (212) 779-6058
www.QuartoKnows.com

First published in 2018 by Motorbooks, an imprint of The Quarto Group,
100 Cummings Center, Suite 265-D, Beverly, MA 01915, USA.

Crestline titles are also available at discount for retail, whole sale, promotional, and bulk purchase. For details, contact the Special Sales Manager by email at specialsales@quarto.com or by mail at The Quarto Group, Attn: Special Sales Manager, 100 Cummings Center, Suite 265-D, Beverly, MA 01915, USA.

10 9 8 7 6 5 4 3 2 1

ISBN: 978-0-7858-3878-4

Acquiring Editor: Darwin Holmstrom
Project Manager: Jordan Wiklund
Art Direction: Cindy Samargia Laun
Cover Design: Brad Norr
Page Design and Layout: Silverglass Design

Printed in Singapore COS072020

On the interior case: The 8335R offered rear-lift capacity of 18,700 pounds. This helps explain why the front end carries 20 of Deere's 100-pound "suitcase" weights. Deere specified 420/85R34 tires for the front. *Shutterstock*

On the back cover: Deere's adaptation of the three-point hitch made operations like this simple. Here a Model 40 Standard front pulls its single-hank ripper breaking up compacted soil. *D&CA*

On the frontis: Deere manufactured its industrial Model L from 1942 into early 1947, however very few were assembled during World War II years.

On the title page: Sometimes known as Cotton and Cane tractors, these high-clearance tractors often went directly from the factory to cotton plantations and sugar cane farms from the U.S. Southeast to the Southwest.

Photos credited *D&CA* appear courtesy of Deere & Company Archives. Photos credited *GLWPI* appear courtesy of Gordon Library Worchester Polytechnic Institute. All other photos © Randy Leffingwell.

CONTENTS

The Dain All-Wheel-Drive and the Waterloo Boy

Deere's board of directors was under pressure. Successful farmers were progressive and read a variety of publications, most of which advocated "power farming." Deere's product lineup needed a tractor. Branch managers had appealed to the board long before then to recognize their customers' growing interests. By 1912 more than 50 other firms manufactured self-propelled farm tractors. Farmers across the United States owned and operated an estimated 10,000 gas tractors—and still ran more than 72,000 steam traction engines. While there still were 20.5 million horses and mules on American farms, Deere's salesmen lost implement sales to their competitors where farmers found tools they wanted hooked up to tractors they could buy. Reluctantly, the board accepted what their dealers saw as inevitable. On July 1, 1912, the group assigned staff engineer C. H. Melvin to produce a tractor.

1912 Melvin. C. H. Melvin's reversible tractor prototype pulled plows when the engine led or towed trailers when the tractor headed the other way. *D&CA*

Theophilus "Theo" Brown, Deere's superintendent at its Marseilles spreader works in East Moline, watched the development. Brown had a mechanical engineering degree from Worcester [Mass.] Polytechnic Institute and joined Deere in late 1911 when he was 32. By 1916, he headed the Plow Works experimental department. He worked closely with Deere Chief Engineer Max Sklovsky. Brown understood Deere's wider purpose for its tractor, believing its success, "would be enhanced if not assured, were it possible to divorce the tractor from the plow and to thus make it available for general purposes," as he wrote in his daily diary.

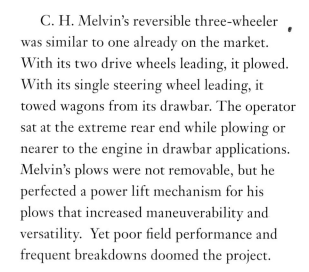

1914 Waterloo Boy Model R-2. Waterloo Engine Company manufactured about 9,310 of the R-series tractors. Its horizontal twin-cylinder engine produced 25 horsepower.

C. H. Melvin's reversible three-wheeler was similar to one already on the market. With its two drive wheels leading, it plowed. With its single steering wheel leading, it towed wagons from its drawbar. The operator sat at the extreme rear end while plowing or nearer to the engine in drawbar applications. Melvin's plows were not removable, but he perfected a power lift mechanism for his plows that increased maneuverability and versatility. Yet poor field performance and frequent breakdowns doomed the project.

DEERE'S HEIRS TAKE OVER THE BUSINESS

Throughout this time, John Deere's son, Charles, ran the company, strengthening it by acquiring his former outside suppliers in order to thwart International Harvester's ambitions. After Charles Deere died in late October 1907, William Butterworth, his successor, further consolidated the company's far-flung agreements. He acquired Dain Manufacturing, a well-regarded hay harvesting and handling tools maker. Joseph Dain, Sr., joined Deere on the last day of October 1910 as a board member when the company acquired his company. Dain's position as a corporate vice president and his fascination with machines drove his interest in tractors.

Despite Melvin's failure, the board knew the machine was a necessary evil. It assigned Joe Dain to study the engineering requirements and the market interest in a machine Deere could sell for $700. Joe Dain soon earned another advocate, Charles Webber, John Deere's grandson. Dain worked

1914 Waterloo Boy Model R-2. This big machine weighed about 6,200 pounds and was 142 inches long. Waterloo sold this model for $850.

through the winter adapting ideas from other makers, including some from his predecessor C. H. Melvin. Dain built a prototype tricycle with a four-cylinder Waukesha engine. Unlike Melvin's tractor where two wheels took the power, Joe Dain used drive shafts and chains to drive all three wheels. His two front wheels straddled the engine, pulling and steering the tractor. His wide, single, center-mounted rear wheel supported implement weight and transferred engine traction to the ground.

Farming was a risky business and that made tractor manufacture risky as well. The conditions affecting agriculture were mercurial; fortunes had been lost in a single bad season. Bad weather followed by bad weather could doom a manufacturer who had sold products on credit.

Yet Charles Webber and Joe Dain watched competitors International Harvester Corporation (IHC) and J. I. Case companies. They felt sure Deere would suffer if it did not press ahead. Board members advocated more testing and development and adding to the number of testing prototypes. Regional branch managers continued their pleas for the new machine.

Board member George W. Mixter, a John Deere great-grandson and vice-president of manufacturing, echoed their concern in a letter to chairman Butterworth. He wrote, "If it be possible to build a small tractor that will really stand up for five or more years' work on the farm, I believe they will be a permanent requirement of the American

farmer and especially in view of the plow trade they carry with them, this possibility cannot be overlooked by Deere & Company."

Dain's first prototype weighed nearly 3,800 pounds and used his innovative friction-drive transmission. This allowed shifting gears from low to high speed while moving. This transmission incorporated a double clutch with faces on both sides. Each clutch had an inner and an outer shaft. This kept both sets of gears in mesh constantly. Engaging one clutch disengaged the other. In a shop test on a kind of chassis dynamometer, Dain recorded a steady 5,000-pound drawbar pull in low gear. In the spring, this proved to be more than 3,000 pounds in actual field tests. Chains drove the axles from the final drive and required upgrading, as did gears, ratchets, and sprockets, to eliminate chances of failure.

The second developmental tractor weighed nearly 4,000 pounds. At Charles Webber's recommendation, engineering sent it to Winnebago, in south-central Minnesota, for testing. Deere's field-testing staff developed its first economic figures while the tractor worked there.

"It plowed eighty acres", Dain revealed to the board, "at a cost of fifty-nine cents per acre, counting the man's time at thirty cents an hour. The soil here was of heavy black gumbo and was in poor condition owing to the almost continual rain. We pulled three fourteen inch bottoms six inches deep at two-and-a-half miles per hour."

Dain had replaced his previous friction-type transmission. This new gear-type version still had a few bugs, but Mixter and Dain believed they quickly could work out these problems in order to get the

1916 Waterloo Boy Model R #1460. By 1916, Waterloo engineers had relocated the radiator to the left side and fitted a large vertical fuel tank. This was the 60th Model R assembled in that year.

tractor to manufacture. A nearly flawless test in Texas cinched the deal. The board approved assembly of 10 more Dain tricycles in March 1916, paving the way for production.

By mid-June 1916, Theo Brown's experimental staff at Marseilles had begun work on the next five Dain development tractors. Engines became a question. Joe Dain had experienced difficulties in testing the various powerplants available on the market. These problems related to engine service and to the inaccessibility of parts (meaning both the difficulty of removing them and the availability of spares). He also had doubts about their general reliability. Dain enlisted Walter McVicker's help.

McVicker designed a new engine specifically for Deere's tractor. Dain advised the executive committee that he expected to receive McVicker's first prototypes in the fall. Meanwhile, Dain installed Waukesha engines into the six new prototypes to go out for testing in August.

Mixter sent a memo to the board on July 13 that fulfilled another requirement of its earlier resolution. He reported that prototype cost figures ranged between $736 and $761, depending on thickness of the frame steel. Adding in $200 for each engine, he estimated cost of manufacture at $600 per tractor.

"This means," Mixter added, "judged in light of other goods of our manufacture that the farmer should pay $1,200 for the machine. This is somewhat higher than had been considered admissible for a three-plow tractor. It is the writer's belief however, that an all-wheel drive will ultimately be the tractor the farmer will pay for."

MORE TRACTOR IDEAS, NOT FEWER

Board members urged chairman Butterworth to consider developing a tractor with more power. Could they produce a machine capable of pulling more than just three plows? Board members wondered if they should plan a motor cultivator as well.

Walter Silver, an engineer working for Brown, had begun developing a motor cultivator early in 1916. His concept mounted an existing horse-drawn Deere cultivator ahead of a powered unit. Operators steered the cultivator by turning a large horizontal wheel that pivoted the front half of the machine. It was an early form of articulation. With the cultivators just below and ahead of them, operators could see the work they did and make subtle adjustments in steering. Motor cultivators were on every manufacturer's mind. International Harvester was developing one, as was J. I. Case Plow Works of Racine, Wisconsin, and several other makers.

While Silver had conceived his own machine as a cultivator, it didn't take long before Brown and others began testing it successfully under a wider variety of field chores. Deere's sales department had enough confidence to publicize the new machine. On the first day of spring, *Farm Implement News* published a story announcing, "Deere to Make Small Tractor."

"Deere & Co., Moline, ILL., are reported to have practically perfected a motor cultivator, unique in principle and design, and which, in addition to cultivating row crops, will do substantially all of the farm work ordinarily done by one team. . . . It pulls as well as pushes,

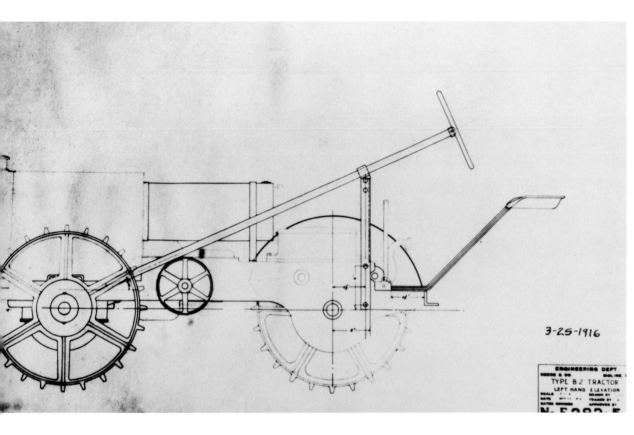

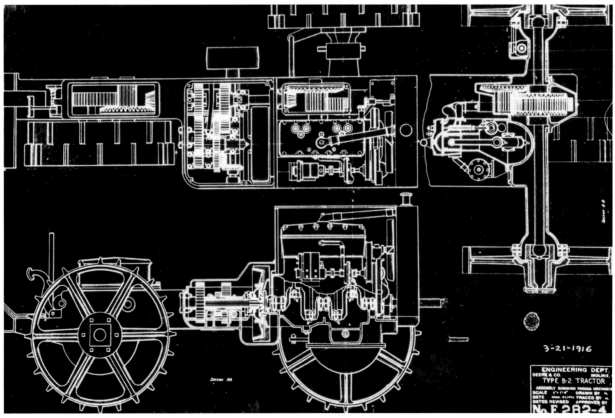

LEFT: 1916 Sklovsky B-2. This is the second generation Max Sklovsky all-wheel drive tractor, the B-2. His choice of a non-Deere Northway engine made it too expensive for production. *D&CA*

BELOW: 1916 Sklovsky B-2. Sklovsky included universal joints for steering. But without a differential, inside and outside wheels always ran at the same speed, making turning difficult. *D&CA*

1916 Dain. This is likely Joe Dain's second all-wheel drive prototype, pulling two reapers. Max Sklovsky's prototypes looked similar. *D&CA*

hauls a drag harrow, small grain drill, mower or single plow; in short, anything that can be done with one team.

"The retail price will approximate the cost of a span of good horses. The trade name of the machine will be the 'John Deere One Team Tractor.'"

WORLD WAR I INTRUDES

Three weeks later, however, the outside world moved in on Deere & Co. On April 4, the US Senate voted for war against Germany, and on April 5, Good Friday, the United States declared war.

Board enthusiasm remained divided over tractor manufacture, between its reluctant chairman and his followers, who were concerned over development costs and sales risks, and those who believed the company was at risk if it did not move ahead. Butterworth had to miss a mid-September 1917 board meeting and asked another member to serve as his proxy.

Henry Ford had entered the market with his small Fordson. Butterworth feared Ford's nearly limitless resources and wrote to his surrogate. "I want it plainly understood that I am and will remain opposed to our taking up the manufacture of tractors and will take steps to stop it if an attempt is made to start. . . . If it comes up, I want you to stop it."

Butterworth saw Ford as a considerable threat to be avoided. Others felt that Ford (and

International Harvester) presented competition that no longer should be ignored. Butterworth learned a painful lesson: Never miss a board meeting, as this one authorized as many as 100 more Dain tractors for manufacture at Deere's Marseille's Works in East Moline, Illinois.

Testing and development continued and brought further success and first sales. In 1916, Deere's Minneapolis branch had loaned a Dain prototype to its South Dakota agent, F.R. Brumwell, who owned three ranches. After a year of use, Deere engineers overhauled the tractor and replaced the original Waukesha powerplant with McVicker's new four-cylinder engine. Brumwell had been impressed enough with the tractor that he had bought three, each with the new McVicker engine. The agent kept one. He sold the other two to customers.

Servicing had been a significant consideration in McVicker's engine design. His concept permitted operators to perform a major engine overhaul, or to remove cylinder head and even pistons and connecting rods, without tearing the entire engine down. Farmers could service the transmission merely by removing a cover plate.

Joseph Dain, Sr., never lived to see his 100 tractors assembled. Overworked and exhausted before his South Dakota trip, he contracted pneumonia and died on October 31, 1917. The board asked his son, Joseph, Jr., to continue his father's work.

ALL-WHEEL DRIVE IN PRODUCTION

Deere's board found a new man to assume responsibility for manufacturing the John Deere all-wheel drive tractors. Elmer

McCormick was superintendent at the Tenth Street factory in East Moline. McCormick, who trained as an engine designer, had worked closely with Dain almost from the start while he studied engineering at University of Illinois. McCormick had contributed in minor ways to the design and engineering of the All-Wheel Drive.

By mid-December 1917, Marseilles was ready to begin assembling the 100 tractors. Manufacturing had let contracts to McVicker for the engines. Deere expected McCormick to complete the first half of the planned production, ready for shipment by June 1, 1918.

As the war raged on in Europe, Deere filled its production capacity manufacturing "escort wagons," ammunition carts, machine gun carriages, and a "combat wagon" for the US Army. In August, the war department ordered 5,000 of these wagons from Deere, the first 166 to be shipped by November 6.

Just after the New Year 1918, the board asked Brown to develop a two-bottom plow suitable for the Fordson tractor. Henry Ford already had shipped the first several thousand units to England's Ministry of Munitions as the "official farm tractor" of wartime England. Now he began delivering back-ordered models to farmers across the United States and Canada. He had assigned his engineers to look into implements, but soon afterward Ford announced that Ford & Sons, his tractor manufacture division, would not produce tools for the tractor. He had decided to leave that task to others. It was a decision that inspired scores of new businesses as well as old, established ones to manufacture implements and accessories for Fordson tractors.

Over the next six months, Brown got involved again with tractors. In late March, he received a new Fordson for the experimental department. A week after that, a new Waterloo Boy tractor appeared unexpectedly at his experimental shops. Brown learned it was intended for use on Deere's experimental farm.

Like many industrial projects, Deere conceived, tested, and produced its revolutionary All-Wheel Drive tractor in utter secrecy. Not only did engineering not want the competition to know, William Butterworth did not want Deere's bankers to find out.

Butterworth had on his hands a superior product full of technical innovation. It appeared to be very reliable and strong. It seemed to be a good value for the money, even at the revised $1,650 asking price.

There still was good reason for Butterworth's caution. Farm and machinery trade journals regularly published stories of the demise of another one or two of the 165 tractor companies still in business by late 1916. North American manufacturers had produced 62,742 tractors in 1917 (and shipped 14,854 to war-torn Europe). However, wartime needs for steel and other manufacturing materials threatened all domestic industry in 1918. The United States Government Priorities Board hinted at rationing steel and limiting first-generation development prototypes. It allowed just 10 of these, no more than 50 second-generation test models, and it held total production by all manufacturers to 315,000 tractors. The government hoped this still would be enough to continue to feed and clothe the world. When the combatants reached an armistice on November 11, 1918, needs for Europe's

restoration stoked manufacturers' optimism. At year's end there were nearly 250 firms claiming they made farm tractors.

All of this provoked young Willard Velie. Willard was Stephen Velie's youngest son, and when his father died, Willard assumed both Stephen's job as Deere's corporate secretary and his father's seat on the board. Even when Willard left Deere to operate Velie Carriage Co. in 1900, he remained on the board. Near the end of January 1918, he urged the board to think bigger. "We cannot profitably make as small a number as 100 tractors, [because] in the process [of doing so] we become competitors of the independent tractor manufacturers, who have been heretofore our 'allies.'" He concluded, "we should build tractors largely and whole-heartedly, or dismiss the tractor matter as inconsequential and immaterial." His comments had almost immediate effect.

Frank Silloway was acting head of sales, filling in for George Peek, who had followed George Mixter into the war effort. Silloway resurrected an idea from earlier board meetings. As Deere geared up production of Dain's tractor, its need for a major tractor manufacturing facility became very clear. Deere already had investigated other producers in the Midwest, among them the Waterloo Gasoline Engine Company, some 110 miles northwest of Moline. Deere already had acquired one of that firm's models and shipped it to Brown for evaluation and possible implement development.

One significant element that made this company appealing to Deere was an acquisition that the Gasoline Engine Company had made back in October 1912, when it purchased the Waterloo Foundry. Prior to that deal, the foundry company manufactured its own tractor, the 8-15 Big Chief Tractor, introduced in 1911. The Big Chief used an opposed two-cylinder, horizontally mounted engine. The Gasoline Engine Company's first tractor, the Waterloo Boy Model 15 carried over the same configuration.

Given Deere's limited tractor production capacity at that time, Silloway had recognized the slow start-up that this insufficient capacity almost certainly guaranteed. Demand could soon outstrip supply, further frustrating his dealers. He wondered if Deere might be better served by acquiring an existing company with established products. He had heard that the Waterloo company was available.

Silloway's first look at Waterloo Gasoline Engine Company impressed him. In 1913, the Waterloo Gasoline Engine Company had introduced its Waterloo Boy One-Man Tractor, a relatively small machine for the time, weighing 9,000 pounds.

The word "Boy" first appeared as a model name on Waterloo gas traction engines in 1896. It's likely the makers of this nearly self-sufficient machine used the name to parody the "water boy" who still was needed to fetch and deliver cooling water to the older-style steam tractors around many farms.

Waterloo produced two models: a single-speed Model R introduced in 1914 and a newer two-speed Model N, released in 1917. Both tractors used a horizontally mounted two-cylinder engine. Both ran on inexpensive kerosene. In 1918, the R sold for $985 and the N sold for $1,150. These came much closer to the price of Ford's Fordson and IHC's International 8-16 than Deere's own All-Wheel-Drive did at $1,650.

1917 Silver Motor Cultivator Number 2. This second-generation Walter Silver Motor Cultivator steered by pivoting the rear tiller wheel. It influenced Theo Brown's thinking on a General Purpose tractor. *D&CA*

Train cars loaded with new Model N tractors, introduced in 1917 with two forward gears and slight updates from the previous R models. This 1920 train carried six Model Ns per car. *D&CA*

Waterloo rated both tractors at 12 horsepower from the drawbar, 25 horsepower from the belt pulley. However, the new N used roller bearings in the engine. What's more, Waterloo had adopted automobile-type steering to replace the Model R's bolster-and-chain steering system. In addition, the Waterloo firm marketed a line of portable and stationery engines capable of burning gasoline or kerosene. Sales of all its models had grown, encouragingly, from 2,762 in 1916 to 4,007 in 1917.

The Waterloo factory was large and well equipped with many new machine and production tools. In addition to the buildings, the company had acquired 38 acres of adjacent vacant land to handle anticipated expansion. With its own foundry, Waterloo Gasoline Engine was "vertically integrated," that is, completely self-sufficient.

Silloway made it clear in his memo to the board that he also liked Waterloo's robust, simple two-cylinder engine. "The tractor, unlike the automobile, must pull hard all the time. There are half as many bearings to adjust on a two-cylinder tractor and half as many valves to grind. There are less parts to get out of order and cause delay. Four cylinders are not necessary on tractors. The fact that a tractor is geared fifty to one instead of four to one eliminates all jerky motion. The engine of a tractor can be made heavy and have a heavy flywheel and can be mounted on a strong rigid frame. Therefore, a two-cylinder engine is satisfactory in a tractor and when it is, why go to the four-cylinder type?" He added forcefully:

"The Waterloo tractor is of a type which the average farmer can buy. . . . We should have a satisfactory tractor at a popular price, and not a high-priced tractor built for the few. Here we have an opportunity to, overnight, step into practically first place in the tractor business."

After a bit of board dithering, Deere acquired Waterloo on March 14, 1918.

Over the years, Brown came to understand the significance of this move. There were options that this acquisition presented for Deere's own tractor plans. With Waterloo's tractors in its lineup, once Marseilles completed the first 100 All-Wheel Drives, they need not go further. Brown's plant finished the last of them in 1919.

When Deere acquired Waterloo, Henry Ford turned out 64 Fordsons a day and every day after in Detroit. Throughout 1917 and into early 1918, he had shipped more than 4,000 tractors to the United Kingdom for its Ministry of Munitions. Selling them at cost-plus-$50, Ford had cut his profit to the bone to aid England's wartime agricultural crisis. But by March 30, he also had filled 5,000 of the 13,000 outstanding back orders on file for his Fordson in North America. Within another three months, Ford doubled his production rate again. By July 1, he was building 131 a day.

The Fordson was not a perfect tractor. But it was small, light, strong, highly maneuverable, and extremely affordable. It weighed 2,710 pounds, and it could pull 2,180 pounds. Ford sold the Fordson for $785. By the end of 1918, he had manufactured 34,167 of them. Deere & Co. had a long way to go to finish second to Ford. Total sales of the Waterloo Boy reached only 5,634 for 1918.

Frank Silloway was not discouraged. Each sale of a Waterloo Boy tractor also meant a potential sale of a Deere plow. Better than that, since Henry Ford had no plans to manufacture his own implements, his 34,000 buyers had to use someone else's plows. Theo Brown and others from Plow Works engineering and Deere's board already had visited Ford to discuss their plows for his Fordson. Deere Plow Works produced a new, 410-pound two-bottom plow, conceived for the smaller tractors, but most specifically for Henry Ford. It weighed 170 pounds less than its nearest competitor.

When the question of distributing Deere plows through Ford dealers came to a vote in mid-September 1918, the Deere board voted to protect its own dealers. If farmers wanted Deere & Co. plows, they could come in and see all the other products their local agent had to offer—including a better tractor!

In 1919, Deere sold 4,015 Waterloo tractors, both Model R and N, down about 1,600 from 1918 sales figures. It discontinued the R at the end of 1919. Tractor contributions to the balance sheet recovered through 1920, when Deere's branches sold a total of 5,045 Model N tractors. The board set production at 30 a day in 1920 and it planned for 40 per day starting in 1921.

Ironically, it was another tractor named Ford that gave Deere and Waterloo Boy their biggest lift. For every legitimate engineer and inventor there was at least one con artist around angling for the farmer's dollar. Some manufacturers sold stock in order to go into business while others only sold stock; that was their business. Some tractors were ill conceived, inadequately tested, inaccurately advertised, and innately unusable. With so many companies in business, name duplication was common: There were three Generals, four Uniteds, five Universals, six Westerns, and 10 outfits named the American Tractor Company. Some schemers used the most legitimate names to dupe some of the nation's least-sophisticated buyers. A clever San Franciscan, William Baer Ewing, leased a plant in Minnesota and hired an accountant named Paul Ford just to make use of Ford's name. The tractor from Minneapolis was a Ford, while, to satisfy his nervous board, Henry's firm in Dearborn incorporated as Ford & Son.

Unfortunately for a Nebraska farmer/state legislator, he bought the wrong Ford. It broke down before it ever reached his farm and never performed as advertised. Frustrated and wary, he bought a second-hand Rumely. In contrast, that machine exceeded its claims and his expectations. This caused him to investigate what else was out there, how well it worked, and what he could do about it. He got together with another Nebraska legislator and the contacted the University of Nebraska. The idea of standardized tests had occurred to professor L. W. Chase, the former head of agricultural engineering, starting with the earliest days of gasoline tractor manufacture.

Each man did his part, writing, introducing, lobbying for, and promoting the Tractor Test Bill. The Nebraska Legislature passed it into law on July 15, 1919. Tests included endurance runs, and engineers determined fuel consumption as well as official horsepower ratings under continuous load. Results were public. The first tractor that appeared for a test was John Deere's Waterloo Boy Model N.

The tests began March 31, 1920, and concluded 10 days later. Several other manufacturers followed, and by the end of April, another seven had begun testing. Before the end of April, Deere & Co. learned that its tractor had exceeded its 12/25 horsepower advertising claims: Nebraska engineers measured 15.98 horsepower on the drawbar, 25.51 horsepower off the pulley. Deere's Waterloo Boy, the first tractor submitted, was the first one certified.

THE PRICE CUT HEARD 'ROUND THE COUNTRY

Henry Ford claimed his goal in bringing out the Fordson was to ease the "producer's" costs so that everyone could benefit. Not only food but also clothing should cost less, he said. The tractor, more efficient and less costly than a team of horses, could aid in achieving that goal. But money was tight. So Ford cut the price of his Fordson. His automobiles were successful, and he used their profits to support losses he could accept to get his tractors onto every farm. On January 27, 1921, Ford dropped the price of his basic Fordson from $785 to $620. Within a month of Ford's price reduction, International Harvester Corporation cut its price from $1,150 to

$1,000. Deere watched, hesitated, and then dropped prices to $890 in July.

Ford waited another six months and then dropped its bomb, reducing his prices another $230. He telegraphed each of his dealers to confirm the new retail price: $395.

In a battered economy, even Ford suffered casualties. Sales fell by 32,000 from 1920 levels. Ford dealers sold only 35,000 Fordsons in 1921. For Deere, it was much worse; they sold just 79 tractors.

If any perspective ever made the Waterloo Boy appear dated, it was seeing Ford's compact little grey tractor on sale at less than half the price of Deere's old-fashioned-looking green and red machine. But Deere's engineers had not been sleeping, nor had Waterloo expected to sell the Model N forever, even if Deere had not come along. Waterloo had begun work on a successor.

Frank Silloway found a compelling secret at Waterloo. The foundry had been one benefit to the acquisition. This surprise played a further role in retiring the All-Wheel-Drive and bringing Deere's tractor in line with its competitors. Once Deere's purchase was final, Waterloo engineers Harry Leavitt and Louis Witry invited Silloway to see the future. They had introduced their Model R in 1914. Over its five-year life, Waterloo had

1918 All-Wheel Drive.
Joe Dain's 1918 three-wheeled tractor utilized all-wheel drive, a not uncommon technology at the time. It weighed nearly 4,600 pounds.

1918 All-Wheel Drive.
Deere introduced the All-Wheel Drive in 1917, with a transmission providing two speeds in forward and reverse. That large single rear wheel was 20 inches wide.

10. By this time, the board felt the Waterloo engineers had the new model pretty close to where it should be.

This machine was smaller than current production models. Between the Waterloo Boy Model N and the last "Style C" prototype, Witry and Leavitt had shortened overall length from 132 inches to 109, accomplished by reversing the horizontal two-cylinder engine on the frame so its crankshaft sat at the rear. They reduced ground clearance. They had shaved weight from 6,200 pounds to just about 4,000. Their improvements in carburetion, intake and exhaust valves, manifolds, and engine bearings increased drawbar horsepower from 16 to 22½, from what basically was the same 6.5 x 7.0 seven-inch engine.

Brown's Experimental Department assembled the next prototype, the first "Style D" tractor. Because it was so complete, development ended after a single example. Deere's competitors had strongly influenced the new machines.

In 1915, Wallis, a part of J. I. Case Plow Works, replaced its inaugural Bear model, a tractor manufactured on a frame, with its new Cub that utilized a unit-frame. This incorporated the engine, transmission, and running gear within one seamless housing. It enclosed moving parts completely, isolating them from weather and the destructive dust of the farm field.

The 1916 Fordson followed suit, although Ford claimed he had patented a version of the unit-frame principle seven years earlier. IHC was ready to introduce its unit-frame International 15-30 about the same time.

While Waterloo developed the A, B, C, and D prototypes, Deere had not been idle. By

sold it in 13 styles, each given an alphabetical designation ascending from RA through RM. They abbreviated the "RN" and it became simply the "N." They restarted the alphabet as they moved into the future, developing and updating a succession of prototypes.

In pre-Deere days, Witry and Leavitt referred to this new model as the "Style A." Deere & Co., now the owners of the new project, liked alphabetical designations, setting in place a pattern that continued for another 35 years. Witry and Leavitt started the second series "Style B" tractors, and assembled and tested another seven prototypes. Style B versions evolved into the more-refined "Style C" tractors, a test model they showed Silloway. He asked Witry and Leavitt to complete two of these by April 1922. They brought Brown in at that point to begin developing implements and tools for the new machine. Deere's board then ordered another late January 1920, Theo Brown successfully had adapted C. H. Melvin's power lift to raise and lower a disc harrow. Six weeks later, Brown signed a patent application for a power take-off (PTO) for the Fordson tractor. At Waterloo, Witry adopted the unit-frame as well. Frank Silloway and others had nicknamed some of Witry's prototypes "bathtub tractors" because of the configuration. Other considerations arose, however. Deere's first production tractor, the All-Wheel Drive, had used McVicker four-cylinder engines. The Fordson and the IHC 15-30 also used fours. But Deere's Waterloo tractors stayed with two-cylinder power. The final decision grew from two

considerations: First, a new engine was costly to design, develop, test, and manufacture, and Deere's board—while it still had large cash reserves even in the tight post World War I economy—balked at authorizing anything new. Second, Deere still had a large stockpile of the Waterloo Model N engines on hand. These fit the Witry/Leavitt prototypes and production versions.

The postwar economy hammered industry. Deere slowed to a three-day workweek through the latter half of 1921. Yet the strained agriculture economy forced the company to adopt a two-day workweek in early 1922, just as Henry Ford cut the price of his Fordson.

1918 All-Wheel Drive. The four-cylinder engine produced 24 horsepower off that 30-inch-diameter belt pulley. It had pressurized water cooling and forced-oil lubrication.

The Model D

It came time to retire the Waterloo Boy N and introduce the new Model D. Through 1921, while Deere had sold just 79 tractors, it had manufactured 786; in 1922, the company only assembled 307. Yet some of the still-divided board pressed hard for first-year production of 1,000 of the new models, arguing the national demand for tractors already existed and when a company offered "a suitable tractor . . . built at a reasonable price to the consumer, it can be sold." Now Deere had such a machine.

Through the end of 1923 and throughout 1924, Deere manufactured 880 Model Ds. While the tractor division showed losses in 1923 and 1924, it turned a profit in 1925 and sales grew from there.

The first 50 Model D tractors appeared with a welded front axle and left-side steering. Witry and Leavitt decided that operators could spin the flywheel for starting instead of a crank as they had on the Waterloo Boys and most other tractors. Deere used a six-spoke, 26-inch-diameter cast flywheel that bolted onto the drive shaft. Witry and Leavitt not only had reversed the Waterloo

1924 Model D. The gasoline/kerosene-fired horizontal twin-engine developed 22.5 horsepower at the drawbar. Designers conceived the spoked 26-inch flywheel as the engine's starting crank.

Model N engine, but they also had improved the horizontal twin, although it remained coupled to a two-speed transmission.

Theo Brown had dedicated several years to standardizing tools and techniques throughout mechanized agriculture. He had reduced the number of tools that operators needed to repair most implements from 24 different sizes and styles to just four. He came into Model D development late in the process. It was not until early June 1924, several months after Deere introduced the tractors, that he spent his first several days with a Model D and the implements meant for the tractor.

Meanwhile, Witry, Leavitt, and Brown had observed numerous broken welds in the front axle after operators hit rocks in the fields. The engineers knew a solid casting would be stronger and also would ensure better balance because of its greater weight. Beginning with the 51st model produced, the new front axle was a two-piece casting. Lightweight, short-wheelbase tractors such as the Fordson and the new Model D sometimes lifted their front wheels far off the ground if the plow caught while the engine was running hard. (Farm newspapers periodically reported that Fordson farmers had been injured or killed by the "rearing tractors," yet many other manufacturers also experienced similar horrors.)

At the beginning of 1925, Deere replaced the original 26-inch-diameter flywheel with a 24-inch-diameter, six-spoke version. The smaller, lighter-weight flywheel allowed slightly higher engine speed. Horsepower output increased with the smaller casting.

1924 John Deere Model D. This was the 10th Model D that Deere manufactured. The first 50 used this composite, welded front axle before Deere switched to a stronger casting.

For all the improvements in farm tractor design and engineering throughout the first 20 years of the twentieth century, it remained a machine for pulling. Up until the mid-1920s, gasoline or kerosene tractors had been best suited to the land and crops between the Mississippi River and the Rocky Mountains.

Brown and others at Deere knew that farmers raising corn in the East and cotton in the South needed to cultivate their rows during the growing season. The tractors Deere and others produced at this time could not do that. Yet things already were changing.

International Harvester Corporation's engineers and board members had used the term "Farmall" privately while referring to prototype cultivators and tractors since November 1919. Soon after legally registering the name on July 17, 1923, IHC implement designer Bert Benjamin and Tractor Works chief engineer Ed Johnston completed assembly of 23 development tractors in various configurations and sizes. They called each one a Farmall. Two of these eventually saw 15,000 hours of use but work with all of them confirmed IHC's faith. The IHC board ordered production of 200 Farmalls for the 1924 introductory year, priced at $825 apiece. In the first four months, IHC sold 111 of them. At year's end, the company increased the price to $950. Once regular production began at Rock Island, Illinois, in 1926, profitability soon followed. The Farmall offered a narrow front wheel arrangement and wide rear tread width with high rear axle ground clearance. IHC, primarily an implement company just as Deere was, offered the new tractor with a broad line of implements. Deere saw this innovative Farmall and ordered Brown to design something similar.

BROWN'S ROLE EXPANDS

Deere's board had offered Brown a seat on April 24, 1923. Brown and other board members saw their first Farmall in early December 1924. Five days later the directors in Moline upped tractor production from 3,000 to 3,800 for 1925.

In late June 1925, and again at the end of September, Brown traveled to Iowa State University (ISU) at Ames to meet with engineering faculty and see a new "universal" tractor the group was designing. One innovation registered strongly on him: the idea of three-row cultivation. By early October, Brown integrated all the ideas he had into his own concept for an "all-crop" tractor. His prototypes carried over the Model D's wide front axle and wheel track width that straddled one row. Following ISU's hints, he configured the new All Crop tractor as a three-row machine that set one row outside each wheel and the middle row beneath the tractor.

ONE ERA ENDS, THE NEXT BEGINS

The first Model D production run of 1926 saw a new solid flywheel keyed onto the drive shaft of the tractors. Enough farmers had broken arms and lost hands between the spokes that the board deemed this change essential. The board also approved development money for Brown's All Crop tractor. His next standardization challenge arose with an industry-wide effort to make power take-off shaft size and rotation speed uniform among all manufacturers.

In short order, Brown added a mower to the All Crop, and by May 19, he had the first tractor ready for testing, complete with a cultivator and power lift. "It is possible to drive to the end of the corn, lift the rigs by power, turn around and come back in the adjoining row without stopping," he wrote in his diary that evening.

THE ALL CROP TAKES ON IHC'S FARMALL

Throughout June, Brown ran the All Crop and its three-row cultivator against the two-row Farmall. Deere's new flexible rig performed much better and steered more easily, especially after his staff "put on a Fordson steering wheel, which is larger." Yet Brown still found advantages to the Farmall, especially where its longer wheelbase and wider rear track made the ride over cornrows more comfortable while cross-cultivating.

Barely a month later, on July 23, 1926, Deere's board member for manufacturing,

Charles "Charlie" Wiman, changed Brown's assignment, directing him to devote his time to the All Crop, which, while improving in tests, still was not perfected. By October 21, Brown's redesigned All Crop and cultivator were ready for field tests. Six weeks later, on December 7, he had a PTO gearbox for the machine. By early February, the board had so much confidence in Brown's PTO experiments that it voted to include the device on each new tractor.

As spring lapsed into summer, Wiman worked to elevate the farm tractor's importance and influence within Deere's product lineup. He continually promoted new developments that improved all of Deere's products. He looked very carefully at his tractors' capabilities and limitations. He concluded that both the Model D and the new All Crop machines had plenty of each.

1924 Model D. This new machine was 33 inches shorter than the Waterloo Boy models, at 109 inches. It weighed 4,400 pounds.

3

The Model C Experimental, GP, and Specials

To sort out the problems, the board authorized Brown on Monday, January 31, 1927, to assemble 25 All Crop tractors and implements and train a dozen operators for testing. On Deere's winter test farm in south Texas, Brown's diary noted on April 12 and 13, 1927, "It looks now as though we had the basis of a real machine." Brown's April 26 entry referred to it as the Model C for the first time. The board approved production, and three months later Deere publicly announced the new machine.

A story published in the *Chicago Tribune*, on Sunday, July 17, claimed: *"Deere & Company is putting out this season a few all work tractors, designed for row-crop cultivation and general farm purposes. The tractor, which apparently is still partly in the experimental stage, is said to straddle one row and cultivate three rows."*

1928 Model C. Deere introduced the Model C with a new three-speed transmission in 1928, though the company manufactured only 75.

More significantly, however, the story noted that Deere "now stands second in the farm implement manufacturing business of the world." Deere had passed J. I. Case Threshing Machine Company to take the second spot.

Deere's branch sales staff grew impatient. Farmall production in 1926 exceeded 4,000 units and output in 1927 topped 9,000 tractors. McCormick-Deering and IHC dealers had plenty of machines to sell. Even though the Deere company's finances had improved, Wiman felt tense. Brown's three-row cultivators were a lingering question mark. Tests of prototype Model C tractors indicated they developed less horsepower than Waterloo

had expected. Should Wiman order everything redesigned and launch new testing? That would cost two more years and set Deere back to third or even fourth place in the market.

Another question cried out: Was Deere to stick with its alphabetical heritage? The "Farmall" name had undeniable appeal and it quickly explained what IHC intended the tractor to do. Would the farmer distinguish between the Model D and the Model C? Around the board table, Wiman, Brown, and others heard names such as "Powerfarmer" and "Farmrite." Frank Silloway fielded objections from some salesmen who worried that over the scratchy telephone lines of the

1928 Model C. The Model C—and the Model D by this time—had gone to a solid flywheel after spinning spokes had caused a number of injuries to farmers.

day the letter "C" sounded like "D." Branch or Waterloo plant staff might misunderstand. So it became GP, General Purpose.

Winter testing in California and Texas proved that the new GP compared favorably to IHC's McCormick-Deering 10-20. More encouraging still, by leap year Wednesday, February 29, 1928, Brown's first tricycle GP was being tested. He fitted it with a two-row cultivator with steerable wheels for better maneuverability around the tender young plants.

Wiman and the board were impressed. A month later, after extensive discussion, they voted to construct new manufacturing facilities at Waterloo to increase D production to 100 per day as well as 50 Model Cs each day. Three weeks later, in the *Farm Implement News* on June 28, 1928, Deere launched its General Purpose tractor.

"Deere & Co., have released their long anticipated general purpose tractor, to be known as the Model GP 10-20. The new model pulls two 14-inch bottoms, drives a 22-inch thresher, and plants and cultivates three rows of corn, cotton or potatoes. In cultivating it straddles one row. The GP will handle from 25 to 40 acres a day depending on conditions. The power take-off shaft is somewhat of an

1928 Model C. Theo Brown initially thought three-row cultivation made sense with the front axle of a Model C straddling a row. Farmers disagreed, and Brown went to four rows. *D&CA*

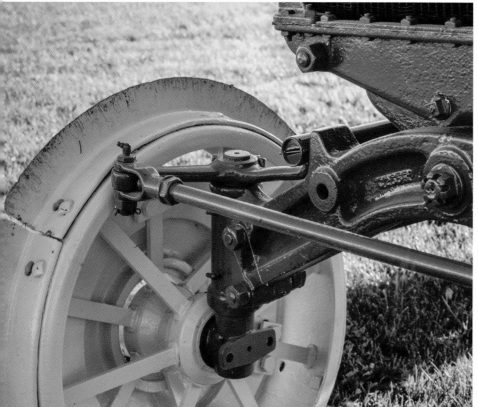

innovation, for it can be operated with either rear or front connections. The shaft turns clockwise at 520 R.P.M. [revolutions per minute,] and has a separate gear shift."

By October, Waterloo manufactured 25 GPs each day. But an old concern reappeared: Deere and its customers expected the GP to produce 25 belt horsepower, yet three weeks worth of production engines had demonstrated only 22.6 horsepower on test benches. The worry went beyond the new GP. For 1928, engineering enlarged the Model D cylinder bore by 0.25 inches, increasing output to 28 horsepower on the drawbar, 36 on the belt. Wiman was concerned because getting much more power for the D required a new engine and then a stronger chassis. He went so far as to propose a new "two-plow" tractor, a Model B, either as a GP replacement or an additional line.

Stories of overworked tractors breaking down came back to haunt Deere. Farmers pointed fingers and accused Deere of poor manufacturing and bad management. Even after Brown's trials around Moline and on Texas test farms, GPs were failing. Some had required rebuilds in the fields. Equally damaging, farmers viewed as unwanted oddities the three-row cultivator and other implements that Brown had conceived for the machine and that Deere had put into production.

What's more, operators found the steering difficult and inaccurate. Most GPs used a steering shaft that ran up the side of the engine.

Road travel was a challenge as the flexible linkages let the tractor drift left and right. Brown and his design team returned to the drawing boards. In a one-year crash program, they produced a Farmall clone, the GP Wide Tread. Its narrowed front axle fit between two rows. Its rear, widened like the Farmall's, straddled them. Brown and his engineers worked hard through the winter and spring on three-row and four-row cultivators and on a new tricycle tractor, completing 23 prototype GP tricycles by mid-April 1929. Charlie Wiman was so confident about the wide-tread model that he authorized Waterloo production even before Brown ran his first field test.

Deere had its GP-WT available in time for spring planting. It met immediate acceptance in the South and Midwest. Wiman's gamble

OPPOSITE TOP: 1928 Model GP Prototype. Deere and other manufacturers concluded two- and four-row cultivation was most efficient. Brown's work on a general-purpose tractor led to this four-row experimental rig. *D&CA*

OPPOSITE BOTTOM: 1928 Model C. Deere's concept for the Model C was general purpose farming, including cultivating in young crops. Front and rear axles provided higher ground clearance.

BELOW: 1928 Model D Industrial. The Waterloo factory first adapted tractors to its freight-moving needs in 1925. This would have run on solid rubber tires in those days.

paid off. This kept Deere in second place in the implement industry. Yet Wiman and Brown had luck, too. Henry Ford, who had declared war on all tractor competitors, had surrendered. The shortcomings of his little Fordson, and his refusal to improve or upgrade the simple machine, finally overcame price advantages. By the end of 1928, Ford moved tractor production to Cork, Ireland, and abandoned the business in the United States. Deere's tractor operation assumed second place as well, runner-up to International Harvester.

It became a vicious circle. Just as tractors became more general in purpose, all-crop in execution, branch sales people, implement dealers, and farmers devised special requirements that narrowed the uses for tractors.

Of the 23 original GP Wide Tread tricycles, Brown's engineers assembled about half-a-dozen

ABOVE: **1928 Model GP.** The GP quickly replaced the C in 1928. For work in dry fields and brush, Deere offered a "flash-suppressor" muffler, seldom seen but fitted to this engine.

LEFT: **1928 Model GP.** This GP carries a Deere No.1 14-inch two-way plow, beneficial for side-hill work or for reversing direction at the end of one row to begin the next.

ABOVE: 1929 Model C. Early tractor farmers pulled implements behind them as horse teams did. By the time this Model C appeared Brown was devising ways to mount the tools. *D&CA*

RIGHT: By the end of April 1928, Brown had a clear concept of the General Purpose–Wide Tread tractor he had in mind. Deere introduced the GP-WT in 1929. *GLWPI*

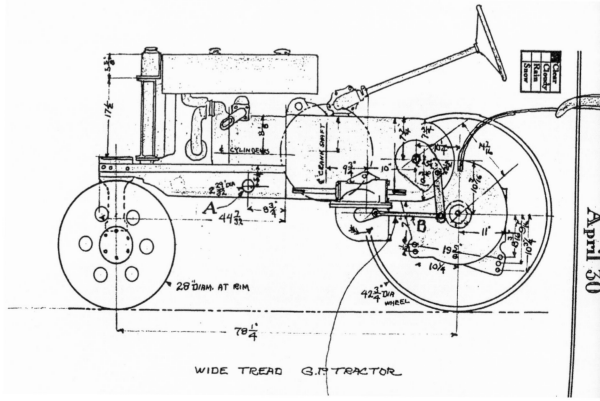

WIDE TREAD G.P. TRACTOR

with eight-inch rear wheel rims instead of 10s. They tapped square holes into the steel. They narrowed the rear track. They did this to meet the needs of potato growers in Maine. Following six prototypes, the factory approved assembling something like 150 "Series P" tractors per year. Barely 14 months later, they discontinued it, instead fitting a Wide Tread tractor with special offset wheels to provide the same tread as the P. Then, for Michigan bean farmers, they created tractors suitable for working 28-inch-wide rows, with different front axles, reversed rear wheels as well as a swinging operator's seat for better viewing in cultivating and harvesting. For Californians, they widened the rear track to 56 inches. Beet farmers worked with 20-, 22-, or 24-inch

rows, and Deere accommodated them. They estimated demand at 350 tractors.

That guess proved wildly optimistic because on October 29, 1929, the New York Stock Exchange crashed. The investment bubble fueled by the beginning of World War I finally had burst. Its immediate impact devastated the financial capitals of the United States and Europe. It took two years more for its worst effects to move into America's farmland. In New York City, stockbrokers jumped out of office windows to their deaths that day. In the Midwest and across the rest of the country, farmers and others perished slowly, or they barely subsisted. Ironically, many of them lived on beans for half a year or longer.

1929 Model GP-WT. Theo Brown developed the WTs, or "tricycles," for two-row farming. The letters stand for "wide tread" or wide track, the distance between rear wheels.

Still, Deere's board understood that even disasters don't last forever. In 1931, Brown and the others answered a long-standing operator's complaint about the GP and the D. They relocated the steering gear and wheel to the tractor's right side. They also increased the D's engine governor speed from 800 to 900 rpm.

"To provide better steering characteristics on this tractor, the steering arms have been re-designed to locate the tie rod behind axle and drag link to equalize the action of steering wheel each side of straight position," they announced. And they continued innovating, improving, and offering updates to ever-fewer customers. A decision, #4337, dated March 2, 1933, at about the deepest point of the Depression, offered both GP orchard models and AC890 Bean Tractors on Firestone low-pressure rubber tires.

ABOVE: **1929 Model GP-WT.** This GP is fitted with Deere's Model GP201 two-row Combined Cotton and Corn Planter. It had the ability to drop seeds in precise intervals.

RIGHT: **1929 Model GP-WT.** The wide-tread or tricycle configuration is easy to see in this view. The rear wheels are intended for the sandy soils of Texas and Arizona cotton farms.

RIGHT: Brown conceived a way to make the front wheels of a GP tractor adaptable to standard tractor dimensions. *GLWPI*

FAR RIGHT: Few of Brown's concepts were simple. Here he devised an onboard hydraulics lift system for front- and rear-mounted cultivators. *GLWPI*

BELOW: Brown's "automatic" rear track adjuster had operators block the rear wheels from moving. The tractor in forward or reverse gear winds the hub in or out along the geared axle. *GLWPI*

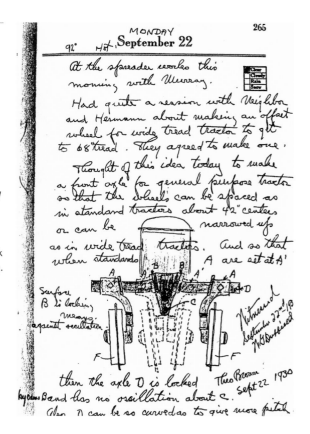

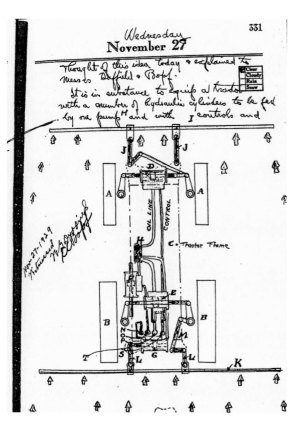

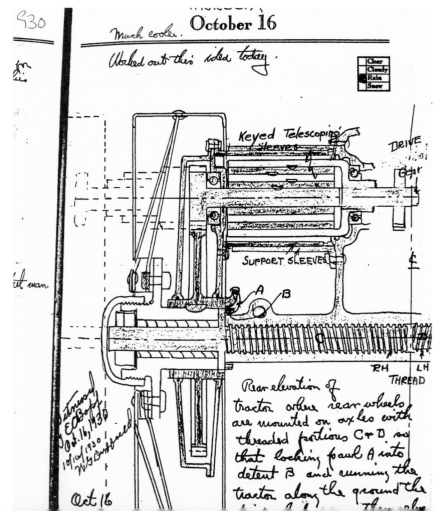

A DIFFERENT TAKE ON WHEELS AND TIRES

About 2,000 miles west of Michigan's bean growers, and further from Maine's potato farmers, another customer, in Yakima, Washington, wanted his tractors without any wheels at all.

In 1920, Jesse Lindeman was barely 20 years old when he moved from western Iowa to central Washington. Two years later, when his brother Harry turned 20, they started Lindeman Power Equipment Company, selling Holt crawlers and harvesters in the Yakima. In 1925, when C. L. Best Gas Traction Company and Holt Manufacturing came together under the name Caterpillar Tractor Co., the Lindemans missed the dealer cut. They quickly picked up a Cleveland Tractor Co. franchise to handle its "Cletrac" models.

Crawlers were a near-necessity in the sandy, hilly terrain where the prevalent crop was tree-grown fruit. Lindeman grew into one Washington's largest dealers.

In 1930, Jesse Lindeman became a full-line Deere dealer. He had lost his brother Harry due to a fatal auto accident, but now his youngest brother Ross joined him. Cletrac's owner had begun "tinkering" with his company, and its crawlers slipped in quality. Deere's Model D impressed the Lindemans very much.

"What struck us," Jesse Lindeman recalled shortly before his death in 1992, "was that here was this wheel tractor, this Model D, and this engine burned what we called 'stove top', this fuel that cost six-and-a-half cents a gallon, no tax. And all these farmers out here wanted that, but they had to have crawler tracks on it. So we just looked in our warehouse and found a used set of Best Thirty tracks and rollers. It was a simple enough thing to do, but it was ugly!"

Another Lindeman brother, Joe, did the tractor testing. The D-crawler got the attention of other farmers in the region, and the Lindemans assembled two more. Handling the tractor was a challenge. Like the Cletrac, it turned only with the use of track

1924 Model D. This very early Model D–with 26-inch spoked flywheel and simple front axle–took a break from harvesting for this photo. *D&CA*

RIGHT: 1931 Model GP-O. Deere manufactured the first Orchard version GP tractor in April 1931. Its hard rubber tires did not penetrate the soil and damage tree roots as metal grousers did. *D&CA*

BELOW: 1931 Model GP Orchard. Deere introduced Orchard tractors in 1931, ultimately assembling more than 650 of them. The price was about $855.

brakes that served to slow or stop the tracks. Farmers tugged hard on brake levers, fighting the power of the engine. "Track clutches had existed on some of the earlier Best and Holt crawlers, but we hadn't figured out that adaptation quite yet," Joe Lindeman recalled.

Deere's experimental department assembled eight or 10 Model Ds as crawlers in late 1931 and early 1932. But they had similar turning problems; engineering shelved the project, dismantled the prototypes, and used the parts elsewhere.

Deere's Portland branch manager and his sales manager approached Lindeman with a new proposition. Deere was interested in producing an orchard-and-grove version of its new GP. Well aware of the work the Lindeman brothers had accomplished with their D-O

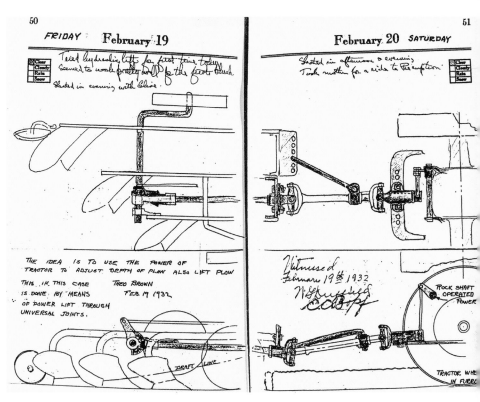

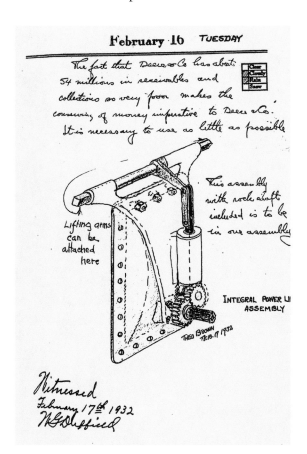

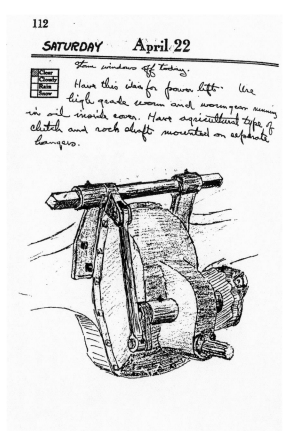

ABOVE: Brown devoted two pages to details of a power implement lift he conceived using PTO and universal joints. *GLWPI*

FAR LEFT: By early 1932, Brown had devised an implement lift system using the PTO and a geared shaft to pivot a rockshaft to raise and lower equipment. *D&CA*

LEFT: Brown continued to brainstorm ideas for the power implement lift. He planned for a "high grade worm and worm gear" inside a cover that bathed the gears in oil. *D&CA*

crawler, the engineering staff in Waterloo wondered—begrudgingly—if Lindeman might do some development work. Jesse learned that "Crawler tractors back there [in the Midwest] was a dirty word. It was something you didn't speak about."

But a prejudice against crawlers was not hard to understand either. While Best and Holt crawlers originated in central California, the new Caterpillar Company relocated to Peoria, Illinois, following its consolidation. Its goal was to be more centrally located to farming and construction needs throughout the entire country. Caterpillar was an aggressive competitor.

Moline shipped a production GP to Yakima. Jesse examined the new tractor, modified its front end, and reversed the rear axle gear clusters. This dropped its overall height nearly seven inches. Now it fit more easily beneath the apple trees growing in central Washington. This tractor steered using track brakes, just as the Lindeman and the experimental factory Ds had done. The differential speed made the outside track turn twice as fast as the braked inside track moved. The tractor actually seemed to go faster through turns. Moline assembled five additional orchard versions based on the Lindeman modifications. Deere shipped them to southern California in mid 1933. Successful results led to regular production.

ABOVE: **1932 Model GP.** Deere's Model GP tractors in this "bean" version were capable of plant cultivation across four rows, as this front-mounted rig demonstrates.

LEFT: **1932 Model GP.** The tractor operator worked on faith with these bean models, because exhaust and air intake stacks obstructed the view of the middle two rows.

Industry-wide, tractor makers were adding gears to their transmissions. For model year 1935, Deere introduced three-speed transmissions for the Model D. Demand for the new Ds surged. That made it easier for Lindeman to get the two-speed GPs. "We didn't even have to change the drive," Jesse explained. "They had a chain drive on the D-O and we had to change it to a smaller chain sprocket. And we put the fenders on it to get under the trees."

He and his brothers converted 24 GPs into GP-O Lindeman crawlers. On some, they fitted long fenders for orchard applications. On others without any fenders at all, they saved weight and costs. They configured the second generation of these tractors with steering clutches that disengaged the drive from the inside track. When the farmer operated the track brake as well, turning the crawler was easier and safer. (Lindeman fitted or refitted all but one of the GP-O crawlers with steering clutches.) The Portland branch watched these developments closely. These improved machines earned new attention from Deere's Plow Works. The Portland branch's managers assisted Lindeman in getting parts he needed, and the brothers let Deere engineers get a

1932 Model GP Orchard. This early orchard model remedied some forward vision problems with its horizontal air intake and downward-turned exhaust. Sheet metal fenders protected low-hanging trees.

1935 Model GP-O.
This GP-O (with steel wheels) mows grass in an orchard. Between 1931 and 1935, Deere manufactured about 600 of these Orchard tractors. *D&CA*

further look at their processes. Deere insisted the crawler be called the Lindeman John Deere; the company also insisted the name Lindeman appeared in the same yellow Deere used for its name. To Jesse, all this was a marketing bonus.

The Portland branch men got sketches and drawings of a smaller new wheeled tractor already in development out to Lindeman, and then an early chassis followed. Lindeman continued the story: "It had no front or rear wheel assemblies. Just drive shafts and axle bolts. We drilled not more than ten holes in that whole tractor to attach our tracks. They didn't plan for any

kind of crawler adaptation when they designed their Model B. It was really just lucky."

American farmers put nearly 230,000 more tractors to work on their farms between 1925 and 1929. The machine population in the United States surged upward from 549,000 to 782,000 during that time, thanks to products such as Deere's Model D and the new GP. According to the United States Department of Agriculture (USDA,) that count reached one million by the end of 1931. Despite the Depression that devastated the nation by then, the USDA estimated that one farmer out of every six owned a tractor. To Deere's board, it signaled growing acceptance of the machine.

To people such as Wiman, it revealed a huge market left untapped once the economy improved: five out of every six farmers didn't yet own one.

Three landmark developments filled the decade from 1924 to 1934. Some historians suggest that IHC's Farmall launched "the Industrial Revolution in Agriculture." Its capabilities catalyzed Deere's engineering efforts. The second landmark, however, came from Deere, and its effect on other manufacturers was equally powerful. Once Brown had perfected the power lift, Deere offered it on the GP starting in 1929. A Works Progress Administration (WPA) study concluded that this one invention saved each farmer 30 minutes every day because operators could pull a lever from their seat, rather than get off the tractor to raise or lower the implement by hand. The WPA suggested that the Power Lift might have saved a total of one million man-hours a year!

The third development came from outside the farm tractor industry. Conflicting reports still add mystery to who deserves credit for introducing pneumatic rubber tires on farm tractors. Some say Allis-Chalmers's President Harry C. Merritt begged some aircraft tires from his friend Harvey Firestone to try on tractors and implements at Merritt's farm. Others insist that hobby-farmer and tire manufacturer Firestone did this first on his Allis-Chalmers-equipped farm and then Firestone force-fed the concept to Merritt. (To further obscure the parentage of the idea, some historians report that Florida orange grove owners mounted truck tires on Farmall tractors in the mid-1920s; steel wheels damaged tree roots. Iowa corn farmers may

have used them as well on their tractors when they towed wagons to town.)

Rubber tires, or "air tires," as farmers called them, significantly improved fuel economy, pulling power, road speed, farmer comfort, and safety. The arrival of pneumatic rubber tires on the farm marked the beginning of the end of horse and mule farming. Farmers could see this turning point starting with the earliest tests in 1930 and 1931. Operator comfort was an important consideration in farmers and manufacturers accepting the new tire technology; Deere's latest overhead worm-and-sector steering system was much more precise, but it had taken all the road-impact-absorbing slack out of the previous linkages. The ride on steel wheels hammered the farmer's arms and shoulders on dirt or on gravel roads. Pneumatic tires cushioned every part of the operator's body. What's more, soft rubber tires allowed farmers to tow their produce wagons into towns with paved streets.

As early as September 1925, Deere had obtained horse-versus-tractor farming cost

1932 "G.X." Tractor Concept. By late June 1932, Brown's diary showed McCormick's ideas for the next tractor, including "over the top" steering for better ground visibility. This became the Model B. *D&CA*

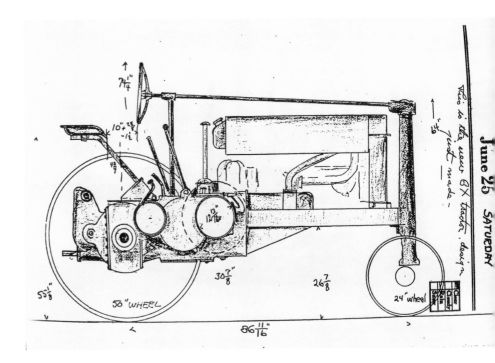

**RIGHT: 1935 Model
GP.** Farmers used the
sweep rake mounted on
the front of this GP to
gather cut and bound
grain shocks left in fields
to dry.

**RIGHT: 1935 Model
GP.** Farmers used the
sweep rake mounted on
the front of this GP to
gather cut and bound
grain shocks left in fields
to dry.

**BELOW: 1935 Model
GP.** Perfecting a
mechanical implement
lift was a significant
Theo Brown engineering
improvement, enabling
this sweep rake to gather
and then carry grain
shocks across a field.

studies from Iowa State University. They
factored in veterinary and blacksmith fees and
calculated the value of time waiting while the
animals rested at the end of each row, plus
time spent hitching and unhitching the team
morning, noon, and night. Figures proved to
Brown that horse farming cost nearly twice
as much per acre as gas tractor farming. The
economics of owning and feeding a five-or-
more-horse team included farmers dedicating
nearly one-quarter of their land to the animals'
feed. Still, horse farming was the only option
for operations smaller than 100 acres in
those days. Wiman was not surprised when
the 1930 Census revealed that four out of
five American farms fit this category. He

recognized that the small, two-plow tractor signaled the beginning of a new way of life for some farmers. His goal was to reach the others who simply never trusted anything new.

Deere's limited tractor lineup put its competitive standing at risk. Three years after the stock market crash, America's economy in 1932 hit bottom. Tractor manufacturer bankruptcies were nearly as common as bank foreclosures and farm auctions. Cotton that had sold for 10 cents per pound in 1930 brought three cents a pound in 1932. Wheat went for 12 cents per bushel and cattle were worth only $4.14 per hundredweight. A Midwestern drought made things worse. Thousands of families lost everything. Yet farmers, then and now, were hard workers who refused to surrender even in the face of indisputable odds. As a result, these same farmers harvested some 371 million acres in 1932, more than ever before in US history.

Those who could survive lived on beans for months, and they had a plentiful supply, for there was no market. Of 186 companies manufacturing or claiming to produce farm tractors in 1922, only 38 remained in 1930. Yet Deere, in business more than 90 years by then, had weathered similar economies.

In late May 1930, Deere shut down its tractor plant; "poor business," was how Brown described the cause in his diary. Yet businesses sometimes got surprised; in early August the board accepted an order for 4,000 tractors for Waterloo to manufacture and ship to Russia. This order reopened the plant. Wiman knew his company had to develop new products even when financial support was tight.

Wiman encouraged Brown to devise new products and new ideas in 1931. In early January, Brown revised the GP tractor to include a live rear axle with reversible wheels and an extended-spline axle. In early February,

1935 Model GP Orchard. Deere records indicate the company manufactured about 30,750 GP models between 1927 and 1935. The arched front axle characterizes the standard track-width versions.

1935 Model GP-O Lindeman. With Deere's blessing, inventor Jesse Lindeman in Yakima, Washington, adapted 25 GP Orchard models to crawler tracks for apple orchard farms in central Washington.

Brown led another meeting to consider "what a new general purpose tractor should be. We are going to follow competitor's outfits carefully in the field this season," he wrote in his diary.

On Friday, March 20, during a directors' quarterly meeting, Brown outlined his requirements for a general purpose tractor, including stability and accessibility (for maintenance and repairs) as well good crop clearance and an unobstructed view for cultivating. He questioned whether the Deere two-cylinder engine was up to the task.

Other engineers, including Elmer McCormick at Waterloo, pushed for more cylinders. Wiman intervened, again citing high development costs, and in late May, Brown got the go-head for a new two-cylinder prototype with his adjustable rear axles. By early July, McCormick had produced Brown's workable prototype. He demonstrated this new machine to Wiman, Brown, and others in Waterloo's test fields. He also showed them competitors' machines for comparison.

"McCormick showed us a modified model of the Wide Tread GP in which the view was wonderfully improved," Brown noted in his diary. "The fuel tank was tapered so as not to interfere with the eye from tractor seat. The seat was raised and brought forward to help see over pulley, the steering put on top of tractor, the air cleaner and muffler taken off the outside and out of the line of vision. His idea is not to make any more wide tread tractors unless the new design."

During another demonstration in mid-September, the engineers devised the idea of making the seat adjustable, both up and down, and forward and back.

GUARDED OPTIMISM AND A NEW TRACTOR

On a cold Tuesday, November 24, 1931, Brown again went to Waterloo to see what McCormick and his staff had been working on. As snow flurries blew around them, they glimpsed the next-generation tractor.

"The new F.X. tractor was assembled and we saw it for the first time," Brown observed. "We were much impressed with its looks. It ran as quietly as a 4 cylinder and without vibration. Weight 4600 pounds. 29 horsepower. . . . We considered making F.X. 36 horsepower and then making another tractor of 24 horsepower on same lines."

McCormick also moved ahead with his G.X. On September 14, 1932, Brown and several Moline engineers and members of the Power Farming Committee went to Waterloo to see McCormick's next prototype run.

"It is a fine looking job, very clean, simple and attractive," Brown wrote that night. "I drove it in the field and it handled very well. Steering was very easy and positive. View was excellent." McCormick told them that the G.X.

weighed 3,702 pounds, compared to the 1932 Model GP at 4,558. Waterloo tested the engine and recorded 24.75 horsepower.

Two weeks after that, on October 13, Brown's Power Farming Committee "went to see balloon rubber tires on a GP with 2 row mounted corn picker. With regular tires and lugs there was not power enough while with the rubber tires conditions as to power were considerably improved. Also the dirt falls off the tires as the tire is distorted." Soon after, Brown got a chance to plow with a tractor mounted on inflatable rubber. "It is surprising," he wrote on October 21, "how much more power is delivered to the drawbar. The tractor wheels don't slip."

STEEL TRACKS JOIN PNEUMATIC RUBBER

As if pneumatic rubber weren't enough of an improvement, on October 27 Brown spent the day in a local field with "Caterpillar attachment for D tractor made by Lindeman of Yakima, Washington. It performed well."

1935 Model GP-O Lindeman. The two tall levers operated track brakes for turning the tractor, which merely slowed one side but seemed to speed up the other.

Brown had assembled quite an audience on the farm outside Moline. Charley Wiman, Frank Silloway, Elmer McCormick, and another dozen board members watched Jesse Lindeman perform his first demonstration.

Brown had spent much of the 1920s working on his own projects. He got the General Purpose tractor designed and built. He and his staff developed and produced planting, cultivating, and harvesting tools for it. Now, Brown observed to himself that the early 1930s were the years in which he watched others introduce new ideas. From McCormick's F. X. and G. X tractors to pneumatic rubber tires, new developments took their first steps during these years. Each of these tantalized or challenged Brown and the other Deere engineers.

On a rainy, snowy November 7, George Nystrom spent the day with Brown. Nystrom worked for Allis-Chalmers. He had experimented with "low pressure tires" for nine months. He told Brown that 20 pounds per square inch (psi) was too much pressure for farming applications; it would allow the tire to slip on the rim. Nystrom concluded that between eight and 15 pounds was the right range. He explained that rubber tires slipped less on side hills than steel wheels did. On lister ridges, the tire flexed to accommodate the ridge. Most significantly, Nystrom informed Brown that Allis-Chalmers was offering pneumatic rubber on its big three-plow-rated tractors as a no-cost option. Brown understood what this decision meant to Deere; he began immediately to modify implements for these more compliant tires.

1935 Model GP-O Lindeman. Lindeman lowered his crawlers to about 45 inches high to slip under fruit trees. This shaved 10 inches off Deere's standard GP models.

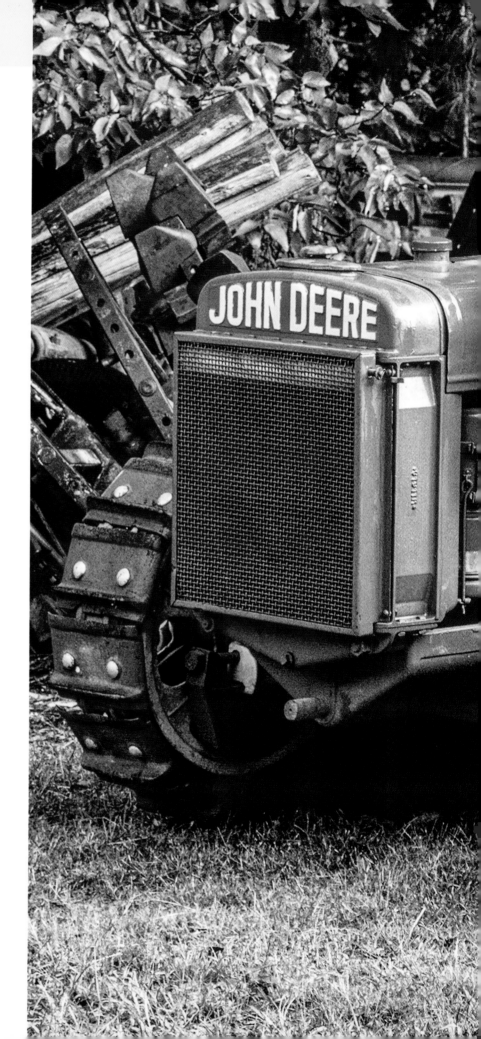

4

Models A and B

In early January 1933, McCormick showed Brown Waterloo's next proposal, the H.X. tractor. "This is a small size job for one sixteen-inch bottom," Brown wrote on January 6, "and is to be as near sixty-five percent of cost and weight of G.X. as possible. It is hoped to get three or four of these tractors built by the middle of April. . . . It may be thought wiser to have this small H.X. precede the G.X. into production. The 7 G.X. tractors at Phoenix and Tucson started work in the field last Tuesday."

The economic news through January and February was the most discouraging yet. Banks in Moline closed for a week or for good. Deere & Company helped bail out one institution by partially underwriting the foundation of its successor, the Moline National Bank.

Operators ran the prototype G.X. during spring 1933 alongside IHC's Farmalls, and they tested the H.X. with Farmall F12s, in Arizona. Deere introduced the G.X. as its Model A. One year later, Franklin Roosevelt's New Deal economy, recharged and jumpstarted, moved

1935 Model A. Engineer Elmer McCormick introduced a separate frame to increase strength, and he tapered the engine cover on the Model A to improve visibility.

RIGHT: 1935 Model A. Theo Brown's mechanical lift relied on cables to raise the shovel on the Model 25 front loader. Gravity brought it down.

BOTTOM LEFT: 1935 Model BN. Model variations proliferated like weeds after Deere introduced the Model A and B tractors. This single front wheel configuration was standard equipment on the Model BN, or B Narrow.

BOTTOM RIGHT: 1935 Model BW. The Model BW, or B Wide, front was at the other end of the front-end spectrum. Deere understood that different crops and soils demanded specific machines.

toward recovery. The board, catching its breath at last, allowed Brown's power farming group the luxury of additional prototypes. Soon seven new H.X. prototypes supplemented ongoing field tests that the seven original evaluation units had completed. Deere needed accurate operations records on each of its prototypes. In January 1934, the company began paying farmers 25 cents a day to write reports as well as plow or cultivate.

On January 30, during the quarterly director's meeting, the board approved manufacture of the H.X. for 1935 as what it called the Model B. Deere had tested the A with three-bottom 14-inch plows and the B with two 14-inch plows. However, knowing farmers often overworked the machines, Deere labeled each tractor for use with one plow fewer: The new Model A was Deere's two-plow tractor and the company designated the B for a single.

ENGINEERING CHALLENGES

Brown and his Moline Plow Works engineering staff constantly struggled to resolve problems others had created. If the implement's hitching point was off the centerline of the tractor, the implement followed behind the tractor at an angle. This was a condition called "side draft," and it plagued many tractors and vexed farmers at the time. Brown designed a new one-piece

1935 Model BW. Elmer McCormick's separate frame on A and on this B model allowed farmers to attach and hang tools from the tractor without risk of structural failure.

1936 Model AI. Once Deere entered the industrial tractor market, it did not leave. It introduced the AI in 1936 and continued production until mid-1941.

transmission housing for Deere's tractors that McCormick used for the A and B models. This relocated both the hitch and the PTO to the tractor's centerline and provided greater ground clearance. Brown designed the casting with a top-mounted access plate through which to replace parts and a bottom-mounted drain plug to facilitate transmission fluid changes.

Tractor designers originally based rear track width on what horse farming had established. This differed for various crops: A two-horse harness placed horse hooves 42 inches apart. So farmers planted and cultivated most row-crops in rows at that width. With

the Model A tractor, Brown's adjustable rear axle offered farmers track width anywhere between 56 inches and 84 inches. Yet, despite the greater track width variability, he and McCormick also improved the tractor's maneuverability through new foot pedal-operated right- and left-differential brakes.

Last but not least, implement engineer Emil Jirsa improved the mechanical-lift feature of the GP. He and Brown developed onboard, self-contained hydraulics. Brown's long mechanical lift lever evolved into Jirsa's short switch. This system raised and lowered the integral implements using hydraulic pull-orpush cylinders. Lift arms attached the tractor to the implement and the farmer adjusted these to set the working depth of front cultivators. Once set, however, the system did not yet offer any automatic adjustment for draft as the tractor rose or fell over hillocks or swales. Deere named Jirsa's system Powr-Trol.

PRODUCTION MODEL A BRINGS VARIATIONS

Brown, McCormick, and their engineers produced eight or so of the G.X./Model A development prototypes. On later prototypes, they tested and perfected Deere's new four-speed gearbox.

Brown's diary noted that the first Model A rolled off the line on March 19, about five weeks later than he had estimated. Full production for a standard two front wheel tricycle began in April 1934. For 1935, Deere introduced the single front tire or narrow front, designated the AN. This accompanied an AW, the adjustable wide-front axle model. The AR, or A Regular, appeared in 1936 as a standard front-axle model. This was a non-row-crop configuration. In 1937, Deere added another letter that specified 40-inch rear wheels instead of the normal size 36-inch equipment. These were the first high-clearance models that Deere offered. The company designated them ANH (or Model A Narrow, High Clearance, for single fronts with a 16-inch tall front tire) or AWH (for the standard or wide front fitted with longer front spindles). Deere offered these high clearance "H" tractors only on pneumatic rubber.

While Deere introduced the orchard Model AO in 1935, the "streamlined" version

AOS appeared in 1937. According to Brown's diary, McCormick, having already narrowed the fuel tank at the rear to improve operator vision, created these next steps at improving appearance along with function. This variation vented the exhaust below the engine and removed the tall air intake extension pipe and fitted modified body panels to protect the trees from the tractor's mechanical works.

The range of Model A tractors grew more diverse. Deere already manufactured a specialized tractor for yards never meant for crops. This was its industrial model, and Waterloo updated it with the AI. A decade earlier, Deere had offered a few Model D tractors with modifications for industrial use. However, this new tractor, introduced along with the AR and AO, adopted features of both models that included faired-in fenders, under-mounted exhaust pipes from the AO, and a modified standard front end from the AR. The AI represented a concerted effort to appeal to the industrial market. This led to a new product line and another tractor philosophy at Deere & Co.

Brown devoted two pages of his diaries to ideas and explanations of a system to raise, lower, and set two separate implements. Deere attorney E. C. Bopf witnessed each page. *GLWPI*

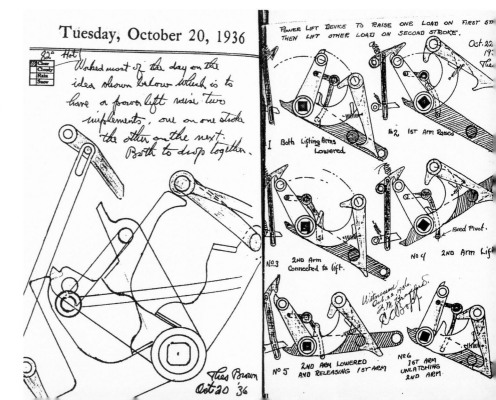

1936 Model AWH.
Brown and McCormick conceived wide-track, high-clearance tractors for farmers growing sugar cane, cotton, and corn. Front and rear axles were adjustable.

Deere first found a need for industrial models in its own factories. With all the heavy equipment that assembly personnel had to relocate and shift around its various plants, engineers improvised a solution to a problem. Steel wheels slipped on concrete factory floors or tore them up, yet solid rubber tires mounted on their steel rims pulled loads easily and quietly.

Ag tractors moved too slowly. To reach useful speeds around the shop and yard pushing or pulling loads of raw materials or implements, engineers reduced Model D final drive gearing. These still provided plenty of low-end tugging power, but they moved the tractors along at 4 or 5 miles per hour (mph)

around plants and nearby roads. Within a year of introducing its agricultural Model D to farmers, Deere had "industrial" Ds running around its facilities in 1925.

By 1926, Deere had introduced advertising materials for "John Deere Industrial Tractors." These offered 40- and 50-inch-diameter rear wheels, extension wheels, and wheel weight sets that ranged from 400 pounds a pair to 1,900 pounds in order to improve traction on smooth surfaces.

Waterloo's chief engineer, Elmer McCormick, was as much an opportunist as an innovator. Working with neighboring Hawkeye Maintainer Company, he and Brown supervised efforts to fit Hawkeye's Motor Patrol road grader to modified Model D tractors with oversize fuel tanks, modified brakes, and a "revised" operator's platform, providing them with fenders or without. Hawkeye produced Motor Patrols from 1929 through 1931.

THE AI AND BI

Brown and his experimental department colleagues continued development work on the GP. McCormick and his Waterloo Tractor Co. engineers and operators put the finishing touches onto the first production Model F.X./A tractors. Waterloo and Moline staffs invested another year testing the Model G.X./B before introducing it as the smaller companion machine. Deere characterized it as being "two-thirds" of a Model A both in power and scale. Moline's improvements to the GP eventually gave it power output

approaching the D. The original performance gap between the two was something that management and the sales staff had found valuable as a sales tool. Now the new B fit the same framework. Deere discontinued the GP soon after releasing the Model B in 1935.

Farmers first saw Deere's AI and BI at the National Road Show. This was an annual multi-company, national dealer's new product introduction event. For 1936, it took place in Cleveland, Ohio, between January 20 and 24. Deere and Caterpillar, partners in marketing and promotion, occupied exhibit space next to each other. At the same time, Deere's Moline branch manager recommended to the board that Deere separate the industrial segment from the agricultural side. In a company as decentralized as Deere was, this was an easy sell.

On the agricultural side, the same evolution of variations occurred with the B as had

1937 Model BI. The BI is compact at just 115 inches long and barely 54 inches wide. Padded seats were an option.

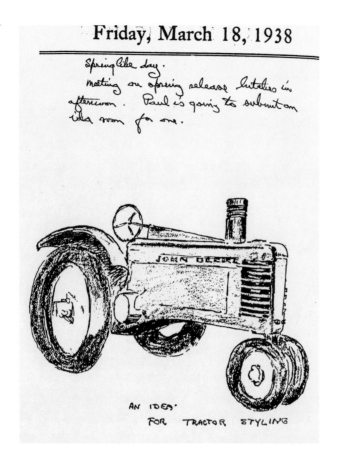

Friday, March 18, 1938

happened with Model A tractors. Soon there were BN models with a single front tire, a BW with its wide front end, the BNH and BWH high-clearance models, BR regulars, BO orchards, and BI industrial tractors. In addition, Deere offered limited-run special versions for beet and other vegetable crop farmers based on their 20-inch-wide rows. These were the BW-40 or BN-40, also available as higher-clearance BWH-40 or BNH-40 models. Designations and variations became as numerous as there were crops to grow in the ground or roads to maintain.

Deere's industrial models appeared in vibrant "Hi-Way Yellow" paint. Caterpillar originally had created this color in 1931, taking on a safety challenge posed by so many companies painting agricultural equipment dark green or grey. Automobile and truck drivers had been seriously hurt at night when they hit darkly painted tractors and other road-building equipment parked on the road's edge. Cat's management pushed it hard as a marketing tool. The vibrant "safety colors" brightened up the crawlers and sparked sales.

However, by the summer of 1937, Deere yielded to its own marketing pressures, offering its industrial tractors painted however the customer requested. Waterloo already finished its own shop tractors bright red. Other customers ordered colors ranging across the rainbow from yellows to blues.

Deere's industrial and agricultural tractors sales improved despite an economy just recovering from the Depression. Deere & Co. had turned the tide against IHC and steadily narrowed the gap between the two companies in sales figures. In early 1935, Theo Brown noted in his diary that Moline

GP tractor production had reached 116 a day and Waterloo's daily output of As and Bs was heading toward 150. Allis-Chalmers, which had introduced its Model U in 1929, had come on strong with rubber-tired versions in 1933 and 1934. In late 1934, A-C released the WC, the first US farm tractor designed specifically to operate on pneumatic rubber tires or "air tires," as farmers called them. At this point, however, all the tractors from all the makers looked pretty similar.

The resemblance was confusing. If it had not been for the beginnings of "logo colors," corporate color schemes, the task of identifying one make of tractor from another across a field might have been nearly impossible. Deere engineers knew this benefitted testing and development work; competitors had trouble identifying new development models from afar. While this helped the engineers keep curious competitors away, it frustrated sales staffs that always looked for something new and exciting to sell.

The company had given its tractors every technological advance it could manage. Deere's board was thinking hard about its customers and the future. In early February 1935, Charley Wiman and his board informed Brown, "that we should immediately begin thinking about the tractor we will build in five years from now. Rubber tires are changing the picture and it may be the tractor should be so designed to do truck service as well as plowing—this might mean spring suspension, and this in turn might mean a device to throw out the spring action when plowing." His next day's notes went further. They listed synchromesh transmissions and overdrive higher speeds for times when tractors served as trucks.

This styled Model B pulls a John Deere Model 5A combine, harvesting wheat. The 5A used either a 10- or 12-foot header. *D&CA*

When, in March 1936, the same group met again, they devised a list of improvements for their next tractor:

- "Consider use of differential brakes to disengage and steer and let front wheels caster around turns
- "Easier attachment of implements
- "More clearance under tractor and lengthen wheelbase of 'B'
- "Better seat—rubber pad or suspension
- "Streamline the tractor."

For some time, McCormick believed that Deere's tractors needed further improvement. It was not an update either he or Brown could provide. It was something called "style."

McCormick wasn't the only one in the tractor business who saw the need for better-looking machines. He was just most vocal. McCormick had annoyed Charley Stone

1937 Model 62.
Deere manufactured the compact Model 62 only in 1937, conceived for the smallest farms that still used two horses or mules. *D&CA*

incessantly with the idea. Stone had headed the Harvester Works since 1923. In 1934, Deere's board promoted him to vice president of all tractor and harvester production. He was a conservative like many on Deere's board and told McCormick that he was not wild about the idea. Nor was he even convinced of the need to do anything like this. However, he declared, if McCormick still wanted to go ahead with it, he wouldn't stop the engineer.

In mid-August 1937, McCormick arrived without an appointment at the fifth floor offices of Henry Dreyfuss Associates at 501 Madison Avenue at Fifty-Second Street in New York City. The secretary and receptionist, Rita Hart, listened to him in polite amazement. Then she rushed into Henry's office and blurted out, "There's a man in a straw hat and

shirt garters out there who says he is from Waterloo, Iowa, and . . . "

"Where?" Dreyfuss said. "Never heard of Waterloo, Iowa."

"He says he is from John Deere and he wants to see you about doing some work."

"Who," Dreyfuss said with growing interest, "is John Deere?"

While McCormick waited, Dreyfuss and Hart paged through the Standard & Poor's *Directory of American Corporations*. Dreyfuss, then 33, had recently landed a choice assignment from New York Central Railroad to redesign its premier cross-country train, The Twentieth Century Limited, so it looked "more like transportation for the Twentieth Century." This assignment had been one of Dreyfuss's big goals. He read more of the description of

Deere's company and recognized his company would be the first industrial designers to work with a big name in this huge industry. He invited McCormick in.

McCormick had traveled 1,100 miles by rail to get to New York. The long ride from Waterloo had given him time to plan his approach. He got right to the point.

"We'd like your help in making our tractors more salable," he said simply.

Henry Dreyfuss was born in 1904 in New York City. He graduated from an elite prep school in 1922 and spent the next year with a pioneering industrial designer doing theater costumes, scenery, and sets. On his own in 1923, he wrote a Broadway theater chain criticizing its set design. For the next five years, Dreyfuss oversaw production of 250 weekly shows running on all of the company's stages. After a break, he began to solicit work as a designer. In 1929, he moved to Fifth Avenue and relabeled himself as an Industrial Designer. His first job was to convince manufacturers that they needed his services. Over several years, his philosophy coalesced. He codified it as his Five Point Formula, emphasizing an object's function: The form of an object should follow *from* its intended use. He urged his clients to consider the object's utility and safety, its ease of maintenance, its cost of production, its sales appeal, and its appearance. The potential of working with Deere & Co. enticed Dreyfuss. McCormick knew he had come to the right place. The two men caught the New York Central and headed west the next day.

What Dreyfuss accomplished is what retired Senior Partner William "Bill"

Purcell described years later as "a clean up really, it wasn't a great change." Purcell knew perspective; he participated in all the "great changes" that came to Deere & Co. over the next three decades. But to the farmer, the appearance and the function was so dramatically improved that somewhere someone coined a word to refer to the effect. Henceforth, those tractors "cleaned up" slightly or changed greatly were forever known as "styled" tractors. Those not bearing an industrial designers' touch still were "unstyled."

The tractors underwent more than "styling" or appearance improvements; there were fundamental design changes. Subsequent tractors received engineering modifications as part of the Dreyfuss industrial design process. But Deere & Co. already had engineers and designers on the payroll so some method had to be designated to explain to the board, the branch sales staffs, and to farmers which of Deere's machines were new and which were not.

Before the end of August 1937, Henry Dreyfuss & Associates had prepared more than a half-dozen design studies for new

1937 Model BI. Deere offered many tire and wheel combinations for the industrials. Electric starters and headlights were optional.

sheet metal for the Model A and B tractors.
A combination of elements taken from two of
them became the landmark "styled" John Deere
tractor. Dreyfuss's team enclosed the steering
column and radiator behind a strong-looking
grille. But the improvements also affected
function because narrowing the radiator cowling
and the gas tank covering significantly improved
visibility ahead and down. Dreyfuss always did
the initial concept sketches himself. From there,
his designers reorganized the instrument panel
to make it easier for the farmer to read while
bouncing through the fields. The back end of
the tractor received some attention as well.
They "cleaned up" its appearance with the goal
of making it easier to recognize the different
functions of various fittings.

The tractor seat seriously concerned Dreyfuss
and his staff.

"The farmer was still sitting down while the
wheel was [positioned] well up there and very
vertical," Bill Purcell explained. "This was because
the rear wheels were still very small and every
bump over a plowed field was so uncomfortable
that the farmer really *had* to stand. So the steering
wheel was still, very logically, set up high.

"When you stood up," he continued, "you
could lean against it, put your hand right on top
of it. In that way, it was very good. However, even
with the tractor at rest, the seat didn't exactly fit.

"Henry asked them once how they designed it,"
Purcell explained. Dreyfuss had carefully developed
his "Human Forms," a thorough collection of
measurements and dimensions for a fictional Joe
and Josephine whose figures allowed Dreyfuss to

1937 Model 62. Deere encouraged its engineers to develop
a very small tractor for the farms still using two horses or
mules. The Model 62 was the result.

design virtually any action or apparatus to fit nearly any size person. The Deere tractor seat did not fit *any* Joe or Josephine. McCormick told them Deere used Pete.

"They looked around the factory to find the fellow with the biggest behind," Purcell continued. "They had him sit in plaster. And that became the seat size."

Deere introduced the newly styled A and B as 1938 models. While farmers regarded these smart-looking machines with caution, the branch sales staffs quickly embraced them as something dramatically "new and improved" to sell. When Deere's advertising brochures and service manuals appeared, the text called them "Tomorrow's tractor today."

Sales of the Model A remained steady at about 11,000 per year from late 1938 through 1942, suggesting that "styling" had

little effect on Deere's big tractor sales. Yet production of the improved Model B (with 20 percent more drawbar horsepower) jumped from 15,000 to 20,000 over the same time. Other improvements sparked sales as well, but Brown's assistant, Wayne Worthington, described the per-tractor costs of improved operator efficiency and modernized appearance as "the best $100 we ever put into a tractor."

"It was a really good looking tractor," Jesse Lindeman said. Decades afterward, Lindeman said it seemed Waterloo had deliberately built Model B's chassis to work with crawler tracks. Lindeman had its own steel foundry and they cast idlers, track rollers, sprockets, and tracks. "But they weren't as nice looking as they should have been," he recalled. "'So I drew up a plan for samples and sent them back to a foundry in Pennsylvania. They

were drop-forge dies, just like Caterpillar's."
Lindeman readily admitted adopting
everyone's good ideas, sampling liberally from
Deere and Caterpillar, including their system
for inserting the connecting track that held
everything together.

Lindeman learned about hydraulics from
Deere & Co., copying what he viewed as
another good idea and adapting it to raise his
bulldozer blade. He previously had used the
belt pulley. He had assembled three crawlers
to demonstrate to the US Army. His operators
raised or lowered the blades on these prototypes
by reversing the pulley direction. Lindeman
developed a three-way blade for the Army tests.
Pulling and resetting a single pin reshaped
the blade as a V-plow, a diagonal blade or a
flat bulldozer. With its PTO-operated winch,
its capabilities exceeded the tractor's compact
dimensions. They didn't earn that contract, but
Lindeman did manufacture a dozen crawlers
on rubber pads during the war. The US Navy
purchased them to run on the paved docks
alongside its ships during re-provisioning. But
the pads suggested other uses to dock crews and
ship hands. Soon they were chugging around
in the holds of ships to clean out iron ore scrap
and slag. The rubber pads avoided cutting up
the ships' steel floors or striking sparks from the
standard steel track pads.

In 1946, Deere & Co. alerted Lindeman
it was dropping the Model BO in another
year. But engineers wondered if he cared to
help devise a crawler version for a new model
they still were developing. Lindeman accepted
the offer, so Deere made Lindeman an even
better one, hoping to buy his company. Deere
acquired Lindeman on January 1, 1947. Deere
stopped production of the BO crawlers after

**1938 and 1939
Model B.** Brown
faithfully recorded the
dramatic difference
between the first-
generation unstyled
Model B and the
second-generation
styled model. *GLWPI*

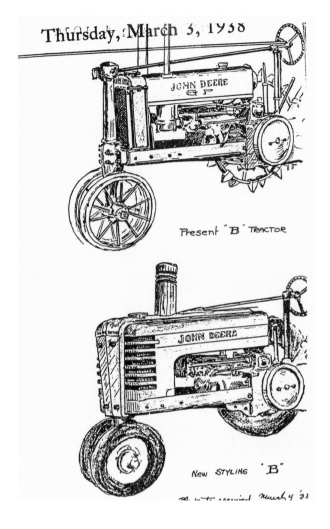

Thursday, March 3, 1938

Present "B" Tractor

New Styling "B"

1,600 units in 1947, when it introduced the
Model M. The MC crawler began regular
manufacture in time for the 1949 season. In
an effort to consolidate far-flung subsidiaries,
Deere then moved all its Yakima operations
back to Dubuque in 1954.

"In the earliest days of these crawlers,"
Lindeman reminisced, "they were the best
way to get around the hills in the orchards.
But then rubber tires came more and more
into use and they got better and better.
That's what did in our small steel tracked
crawlers for orchard use."

Deere's chief engineer, Max Sklovsky,
continued working on his compact single-plow

tractor prototype. It remained flawed in concept
and execution, yet Sklovsky was resistant
to suggestions or ideas for improvement, a
behavior that challenged Brown's patience. It
got worse in mid-May however, when Brown
saw yet another approach to the question of
small, single-plow tractors.

On May 13, Brown got to see Ira Maxon's
prototype Model Y. Where Sklovsky's invention
plagued him, Maxon's logic impressed him.
Maxon was Max Sklovsky's son-in-law, a
relationship that further complicated Brown's
reaction to their two approaches.

Brown and Charley Wiman watched
Maxon's Model Y in the field on June 12.
Brown liked the machine he saw. This created
a dilemma that further unsettled him. He
took his work seriously and had come to
believe that each prototype needed a full set of
implements and tools before its development
could advance. Wiman understood this and
ordered Brown, already overworked, to take a
two-month vacation.

When Brown returned in early September
he saw a Model B with a five-inch-longer
wheelbase, an experimental quick-detachable
mounting apparatus for integral implements,
and the latest variation of the larger three-plow-
rated prototype tractor known internally as the
K.X. Barely two weeks later, the board approved
production of the K.X. as Deere's Model G.
McCormick's engineers at Waterloo already had
begun work on its companion, the O.X.

Throughout 1937, final development work
continued on the Model G. Waterloo geared
up to manufacture the new tractor. Farmer
acceptance of rubber tires led Deere to offer
them extensively, however the decentralized
management meant that one factory did not

always communicate with another. By mid-May, the company needed 76 different tire sizes to fit the tractors, combines, and other implements it manufactured. Brown and McCormick co-organized a wheel committee to simplify and standardize the sizes and numbers of tires and wheels that Deere inventoried, trimming it to 46 varieties. They also began establishing design parameters for a new 12-horsepower general purpose tractor that provided good front end ground clearance.

INDUSTRY RETURNS DESPITE FARMING'S CHALLENGES

Competition outside Deere was fierce. The company held on firmly to its number two position behind International Harvester in tractor, combine, and implement sales. Yet industry observers knew a part of the credit for this was due to Deere's marketing arrangement with Caterpillar. The two companies manufactured essentially non-competitive products and benefitted from shared sales outlets in markets where that made sense.

Work accelerated in experimental departments. In some cases, engineers competed not only against other manufacturers but also against plants and divisions within the same corporation. Decentralized management characterized Deere & Co. beginning in the 1880s and 1890s. William Butterworth had codified this concept into company *modus operandi* in 1915 when he made each factory and every division autonomous. This independence sometimes led to duplicated efforts. Occasionally, however, this resulted in remarkable innovation. At Deere, it had led to the GP tractors from Theo Brown's Plow Works, and to the A, B, G and H models from McCormick's Waterloo Works.

Development continued with the O.X. prototype. In November, engineers began modifying the rear end to accommodate the "Q.D." quick-disconnect integral implement mounting system.

1939 Model B. This line drawing was part of the press information kit introducing the Styled Model B to journalists, dealers, and customers. The dramatic shadowing on the nosepiece emphasizes its strength. *D&CA*

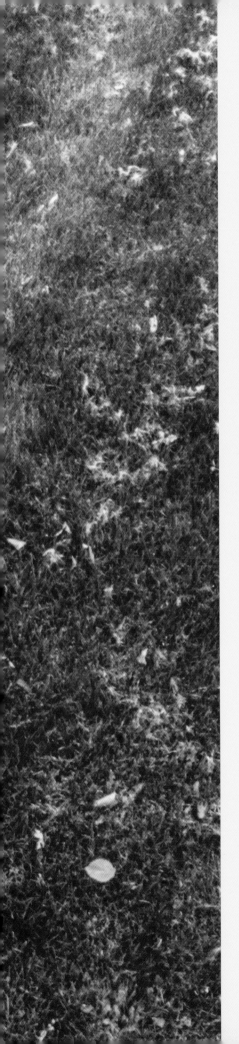

Models L
and LA

John Deere's alphabet has sparked curiosity about what might have been *between* the letters. When Ira Maxon's engineering group first conceived the Model L, they referred to its earliest prototypes as the Model Y. It is a gap and a leap that has not yet been explained. Regrettably, no codebook exists to decipher why the K.X. became the G or how it happened that the "Y" became the Model *number* 62 and why this designation reverted to the alphabetical "L."

Deere management conceived this tractor for the smallest family-run operations, a category that still described most US farms. These farmers used a two-horse or two-mule team for plowing and harvesting. Maxon designed the Y and developed it as a one-plow tractor. Recognizing that this machine necessarily would be smaller than any other in Deere's line, the board chose to veer away slightly from previous procedure. Instead of assigning the new project to Brown's overworked

1938 Model L. This version succeeded the 62 from 1936 into 1938. Deere assembled around 1,500 of these Model L tractors.

1938 Model L. The rear wheel extensions are called mud lugs, designed to give tractors traction through the slipperiest surfaces. The Model L sold for $477.

Plow Works experimental department already developing full implements for the G and H models at this time, or to McCormick's engineers at Waterloo still perfecting the G and H, this request stayed in the Moline Tractor Division, established in May 1941.

The challenge for its chief engineer and manager Maxon was that most of Deere's tractor research and development funds went to Waterloo. These resources were still needed in developing the G and H as successors to the Models A and B. The board felt it was critical to bring them to production on time. Maxon brought in an old friend, Willard Nordenson, a former engineer with Deere who had gone freelancing during the early 1930s. While he was known more as an engine

specialist, Nordenson quickly embraced the "whole product." Although Deere gave Maxon virtually no budget to develop this new machine, Nordenson understood its power requirements were much less than current multi-plow production tractors. With no money available to design a new Deere engine, he went shopping outside.

He selected a two-cylinder eight-horsepower Novo. Nordenson's first few prototype Model Y tractors used the gasoline-only Model C-66 engine. He knew that farmers needed his tractor not only for plowing but also for cultivating. Forward visibility was critical. He mounted the small Novo longitudinally and upright. Its crankshaft ran parallel to the direction of travel unlike the transverse and horizontal arrangements of engines on other Deere tractors. He inserted a Ford Model A automobile transmission with a foot-operated clutch along the drivetrain ahead of the differential. This combination, born of economic necessity, later provided an advertising advantage. The gearshift followed the standard "H" pattern familiar to any car or truck driver. Deere promoted the tractor for use on estates, golf courses, and even cemeteries to operate lawn mowing equipment.

He mounted the engine on a stubby frame consisting of two nearly parallel round tubes joined together by a flat cast-iron plate beneath the radiator. He cast this housing with steel rings in two openings to provide a better weld for the tubular pipes.

The design never stopped delivering challenges. Nearly every other designer of larger tractors before Nordenson had located the front steering wheels ahead of the radiator and aligned the chassis so it ran nearly parallel

to the ground. On the Y, Nordenson placed the front wheels under the engine's rear cylinder, just ahead of the flywheel. This severely shortened the already diminutive wheelbase, greatly enhancing maneuverability for cultivating.

The chassis sloped down noticeably from front to rear by virtue of front spindle heights versus rear axle placement and tire diameter. In addition, because the tractor's overall length was so short, 91 inches compared to the 120.5 inches of the earliest unstyled Model Bs, Nordenson offset the operator's seating position slightly to the right to accommodate all the mechanical components. He also shifted the engine and steering positions slightly off-center to the left as seen when looking at the tractor from the rear. For the first time on a Deere tractor, farmers operated both rear brakes using the right foot to press two pedals linked together. Nordenson kept the foot-operated clutch pedal on the left, just as in cars and trucks.

1941 Model LA. Deere replaced the Model L Hercules with its own vertical twin that developed 13.1 horsepower at the drawbar.

Nordenson's group assembled about two dozen Model Y prototypes, but only a few of these used the Novo. Its single-quart oil capacity was insufficient to keep the bottom end lubricated when the little tractor worked hard. Crankshaft and main bearing failures quickly doomed the Novo. Early into the development of the Y prototypes, Nordenson and his colleagues selected a Hercules two-cylinder 3x4-inch engine. That version, basically produced to a John Deere design, remained in use through the 1937 production life of the Model 62, the second-generation development tractor. It carried on through the 1938 production of the unstyled Ls as well. This amounted to slightly more than 1,580 tractors.

While Henry Dreyfuss had performed only a "clean up" on the Model A and B tractors, the Model L was his virtually from the start. For Deere's larger tractor markets, the company continued offering unstyled versions alongside the styled machines. The Tractor Works introduced the "styled" version of the L on the model's first anniversary in August 1938. Starting with the 1939 model year, Deere no longer offered original unstyled model Ls.

Beginning in 1941, Deere produced its own engine for the styled L successor, the LA. Deere modified the Hercules used in the prototypes to accommodate a generator and self-starter, by this time a common option package.

Immediately before Deere introduced the LA, it ceased production of both the L and the early styled L Industrial. Deere then introduced an LI to replace the earlier model. It fitted these with the Deere engine and the LA's larger rear wheels.

1941 Model LA. Henry Dreyfuss's styling spread throughout Deere's tractor line. This updated L, called the LA, was 2 inches longer and 700 pounds heavier than the L.

6

Models G, H, A, B, Styled D, and M

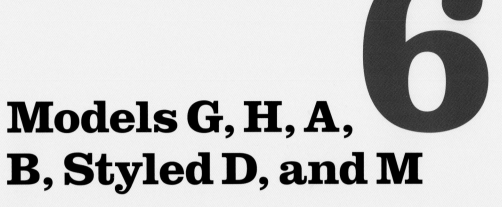

The reality of the late 1930s was framed by experiences from the first half of the decade. While prosperity had returned, few trusted it. Farmers who had gotten by with aging or failing equipment during the early 1930s replaced it in 1935 and 1936. But they bought larger tractors, harvesters, implements, and other tools. For many farmers, this eliminated the need for a hired hand. The US Census concluded that unemployment or underemployment in the cities had kept farmers, their families, and their farmhands on rural lands because they couldn't afford to leave. When rains returned to the Midwest in the late 1930s and farm crop output increased, prices dropped again, creating another depressed cycle that lasted through 1938 and finally abated in 1940 with the onset of the next world war. Then, too, the cities offered millions of jobs with regular hours and days off.

1939 Model H. This row-crop-configured Model H looks smart and modern with its Henry Dreyfuss styling. Dozens of mechanical improvements accompanied the new look.

Farmers had less help and more work to perform. Tractors assumed a greater obligation. Branch managers reported to manufacturers farmer requests: more power, more capacity, more versatility—more, more, more. Early Deere GPs had been two-plow rated; that was no longer enough.

Deere made each successive change an increase in strength and an improvement in features and appearance. By mid-February 1937, all the executives within Deere had seen the first styled Model B tractors. That May, Deere & Co. began production of its new Model G as a three-plow tractor with the four-speed transmission recently introduced on A and B models. It offered the new G on steel wheels or pneumatic rubber. With its three-plow capability, Deere promoted it as the tractor to replace the farmer's five- or six-horse teams. By

1939 Model H. When Deere introduced the H in 1939, it sold for $595. Over six years, the company assembled some 58,000 in several variations.

the time the first production Gs appeared for sale, Deere had recorded its best year ever.

GOOD BUT NOT GREAT

Henry Dreyfuss designers helped with the early Model G tractors. Unstyled at introduction, these models featured a small engine cowl and an exposed, sloping steering column extending to an outside vertical steering post. This slim configuration was possible because petroleum distillates and some of the other lower grade fuels burned cooler than kerosene or gasoline. McCormick's engineering staff utilized an 11-gallon cooling system, expecting to keep the engine within the proper temperature range for its maximum operating efficiency. However, this small radiator was no match for summer

Wednesday, October 4, 1939

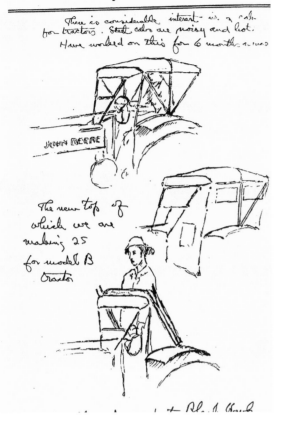

Monday, March 11, 1940

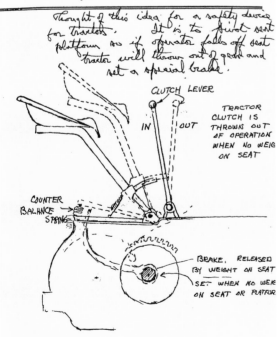

Friday, October 6, 1939

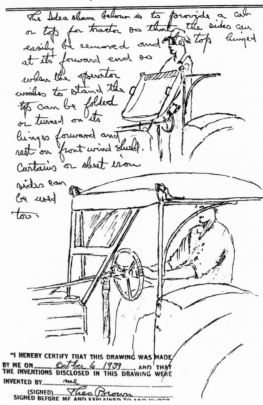

Tuesday, March 21, 1939

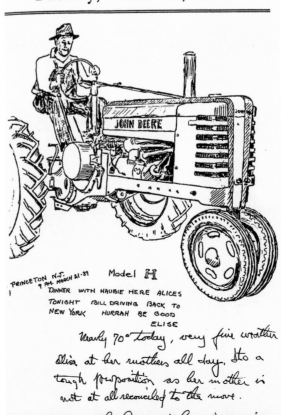

FAR LEFT: 1939 Soft-Top Cab Concept. Noting that metal cabs are hot and noisy, Brown designed a cloth folding top system similar to automobiles, except this one folded forward. *GLWPI*

LEFT: 1939 Hard-Top Cab Concept. This October 6, 1939, concept provides a more durable, hardy top covering than the soft-top Brown conceived two days earlier. *GLWPI*

BELOW LEFT: **1940 Safety Seat.** Brown's idea here is to protect an operator and eliminate the chance of a runaway tractor if the operator falls off the tractor. Levers disengage the clutch and set a brake. *GLWPI*

BELOW RIGHT: **1939 Model H.** It was a given that engineers and designers must be able to draw. More than that, Deere required that these individuals keep daily diaries of their ideas. Brown often added personal records. *D&CA*

TOP: 1940 John Deere Logo. In late February 1940, Brown proposed modifying the company logo for dealers by deleting "Moline Ill" and replacing it with "Quality Farm Equipment." *GLWPI*

BOTTOM: 1941 Model 101. Brown called his Model 101 concept the "all visibility" tractor, a one-row cultivator with an excellent view of the crops. *GLWPI*

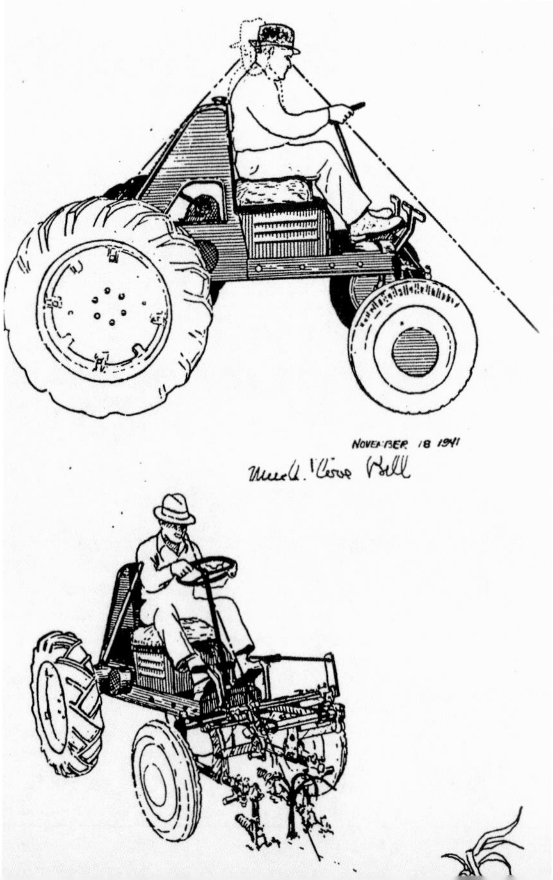

work in warmer climates. More than half of the first-year's Model Gs suffered chronic cooling problems. Later versions incorporated a 13-gallon radiator with a creased top to accommodate the steering wheel shaft.

Farmers heard a different sound with the Model G. Engine noise of each Deere tractor, widely known as "Poppin' Johnnies," varied slightly due to bore, stroke, and engine speed. The big D produced the fullest voice of the lineup. The B's "pop" had a higher pitch due to its engine's higher operating speed. Both the A and G ran at the same engine speed but their bore and stroke differed and this affected the tone of their sounds.

By the time the public saw the Model G, Waterloo engineers were testing their other project, the O.X. companion to the G. Even as they acknowledged the market for larger tractors, they conceived this as a single-plow machine. They intended the O.X./Model H to supplement the agricultural market not quite fully served by the Model L. This was a true row-crop GP, the perfect second tractor for farms needing something suitable for large vegetable gardens. After a series of development models, production of the tractor began in mid-January 1939, built only as a dual-front-wheeled tricycle. On March 21, Brown and others saw the styled Model H, and production changed over shortly afterward.

HENRY FORD RETURNS

Henry Ford discontinued tractor manufacture in the US, but he never forgot the machine. Once his new V-8 was reliable and his next cars were on the test tracks, he put his engineering staff to work devising his next tractor. When

Irish inventor/salesman Harry Ferguson showed up at Ford's doors with a very workable concept, Ford threw his entire tractor engineering staff at the machine. In late April 1939, Henry Ford presented it to the public as the Model 9N "Ford Tractor with Ferguson System." Due to Ferguson's clever engineering, his innovative three-point hitch, and the integral two-bottom plow, the lightweight machine had the effectiveness of something bigger and heavier. With Ferguson's help, Ford offered a full line of implements. Yet, like Deere's early Model G, Ford's 9N had shortcomings. Ferguson's hydraulics, while used in a new configuration to control plow draft, worked perfectly only in ideal conditions.

Meanwhile, Brown worked on cabs for the Model A and B tractors. He disliked steel cabs; they were noisy and hot. Brown's initial creation functioned just like a convertible top on an automobile. A later version flipped over past the windshield when the operator wanted or needed to stand. By late October, Brown had sent a prototype flip top to Waterloo for testing.

All throughout 1939 and into early 1940, Waterloo engineers also worked on a single-front-wheel variation of the Model H, designated the HN, for vegetable farmers. Waterloo's level of development success was so high that McCormick's staff fabricated just one HN prototype. The second HN tractor was the first production model, introduced for 1940.

By late April 1940, Brown's perspective on Ford's 9N tractor had changed, as it had throughout the industry. After Deere's Annual Meeting on April 30, 1940, Brown wrote, "Ford competition is beginning to worry us. The hydraulic control particularly. I have two patent applications pretty well advanced with claims allowed which should be a better way of

1941 Model HNH. One such variation on the Model H was this narrow-front, high-crop tractor. An electric starter and lights were optional on all H series tractors.

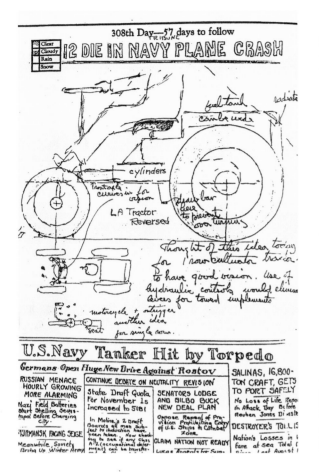

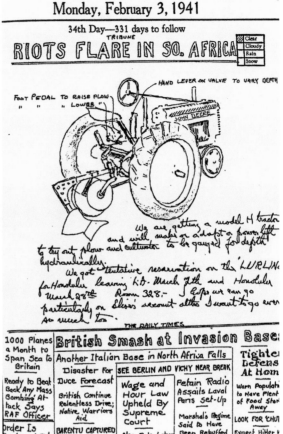

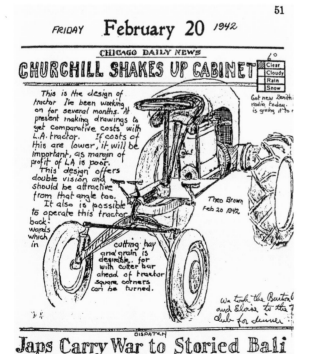

doing what Ford tries to." When Brown visited Waterloo to discuss his new power-lift concept, he saw their new R.X. tractor, one to replace the Model A as soon as July 1941. Brown saw his next assignment: "We will want to work on implements to be controlled hydraulically for depth," he noted in his diary. "Need a device so tool will return to same depth when lowered."

THE NEXT WAR INTRUDES

As early as December 1938, Brown's diaries—always much more than mere documentation of his work and inventions—began reporting on events in Europe. On December 5, he wrote, "Hitler's persecution of the Jews in Germany is a terrible thing and is a

OPPOSITE: 1941 Model HWH. This was the other version of the Model H high-crop tractor, with adjustable wide front track. Extra tall front spindles lifted the strong front axle above young plants.

FAR LEFT: 1941 Model LA Reversed. Brown became obsessed by war news reports, often copying newspaper pages among designs like this small tractor with the operator sitting ahead of the engine. *GLWPI*

LEFT: 1941 Model H. Between headlines of the world, Brown sandwiched his concept for a power lift for the new Model H tractors. He also mused about vacation plans by ship to Honolulu. *GLWPI*

BELOW: 1942 Small Tractor. As the war accelerated, it seemed sometimes Brown interrupted his own war coverage to invent something, in this case another small tractor that he had had in mind for some time. *GLWPI*

RIGHT: 1944 B-O Lindeman. Jesse Lindeman manufactured just 12 of his crawlers on these rubber track pads for a U.S. Navy contract. Rubber did not tear up concrete docks.

BELOW: 1944 Model B-O Lindeman. After the Navy declined further orders, the rubber-pad Lindemans found use on board ships, cleaning out storage holds without causing sparks.

reincarnation of the dark ages." A year later, on Thanksgiving Day, November 30, 1939, he observed, "Russian invasion of Finland is horrible." On March 1, 1940, he noted that "Sumner Welles of the State Department is in Germany to see Hitler and find out if any peace is possible."

The war quickly impacted Moline's factories when government buyers asked for cost estimates to assemble gun carriages. Deere found itself balancing civilian agricultural production against increasing US government contracts with some difficulty. Brown reported that on February 12, 1941, the Waterloo factory produced 321 tractors in three shifts, its highest output to date. The mix of government contracts and agricultural machinery production began to represent significant numbers. By early April 1941, the board reported "Deere & Co. has war contracts for about 18 and a half million dollars for differentials and transmissions in tanks." Days later, the board approved another production variation for Waterloo's Model H, including an HWH tractor similar to AWH and BWN models. Production began about two months later. Soon after, Deere began assembling the HNH narrow version. These two series continued only through mid-1942.

At the end of 1941, Dreyfuss designers and engineers had caught up with the Model G. Deere planned to introduce the new version early the next year. And as the year drew to a close, Brown conceived a small single-row tractor by reversing the direction of travel of the small LA model. His notes for November 13, 1941, described the operator sitting ahead of the engine rather than behind it. He went further in the next few days, pushing his

wartime frugality. He envisioned a simple single-row cultivator that used a motorcycle with an outrigger wheel. But the war upended everyone's plans early one winter morning.

1944 Model B.
Lindeman manufactured BO crawlers from 1939 until 1947. He assembled 1,675 in all.

PEARL HARBOR CHANGES EVERYTHING

The Japanese attack on Pearl Harbor on December 7, 1941, reformulated plans and policies for all American manufacturers for the next three years. For Deere, its styled Model G appeared with a manual starting system as its predecessors did. Electric starters and batteries used precious raw materials. Deere shipped the G on steel wheels and delayed introducing the long-anticipated six-speed transmission. The federal government reasoned that with raw materials and other product shortages

in effect, annual price increases represented an avoidable hardship to domestic consumers. Deere and others were not permitted to raise prices to cover improvements to existing products. They only could charge a higher price for a new model. Changing the designation of a vastly improved, Dreyfuss-styled, Model G to "GM" allowed Deere to recover the costs of all the improvements.

The plans for the 1942 Model G included styled sheet metal work as well as rubber tires, electric lights and self-starter, and the new six-speed transmission. Deere delivered very few that were so fully equipped. At the same time, however, the company introduced the GN (single front) and GW (wide front axle.) Farmers could order rear axles that provided tread width as great as 104 inches.

LEFT: 1946 Model AWH. Deere introduced the styled Model A in late 1940. The high-crop versions raised front and rear axles three inches, providing 24 inches of clearance.

ABOVE: 1945 Model BR. Deere kept its unstyled tractors in production into early 1949. This BR–or "regular" or standard–front end had optional electric lights and starter.

RIGHT: 1946 Model D.
The beefy Model D
was Deere's only styled
tractor not to adopt Henry
Dreyfuss's initial horizontal
grille-and-side-panels
treatment. The styled D
arrived in April 1939.

BELOW: 1946 Model D.
While Deere announced
production end of the
Model D for early 1953,
demand (and remaining
parts inventories) led to
the assembly of nearly
100 more.

KEEPING THE HOME FARMS WORKING

For farmers throughout the US, Brown continued work on his own prototype reversed-and-revised LA, now designated the Model 101. McCormick, as chairman of the Power Farming Committee, reported to the board: "Though this tractor does not in its exterior appearance follow the conventional line, it might be a forerunner of a new design of tractor." Six months later, the 101's radical looks still provoked strong responses. At the April 27, 1943, quarterly directors meeting, Brown showed his new tractor. He recalled, "They all said it was so revolutionary they would have to get used to it but thought it offered real possibilities."

Wiman's insistence on continuing civilian product development during the war kept Deere's engineers and its plants occupied. The company had two projects in the works. Deere meant to introduce the first tractor immediately after the war ended, and as with all the models before, it provided the basis for the transitions that followed. The other, a machine already moving through the design stage, offered a development so significant that it changed the thinking about every tractor that was yet to come.

World War II forced diversity on Deere & Co. Alongside tractors, tank engines, and military crawlers, the company produced artillery shell casings; when the war ended, the US government asked Deere to continue that and other projects as well.

This signaled to Wiman the need to increase capacity. Land around the Waterloo facilities largely was spoken for. Deere already had committed to purchase the space available in Moline. So it looked outside the immediate area. Criteria included "navigable water" on which to bring raw materials in and to ship completed products out. The board selected Dubuque,

1946 Model G. Deere's Model G replaced the ✗ Model A beginning in 1941. This 1947 model sold for $1,879 when new.

✗ NOT TRUE!

ABOVE: 1947 Model LI. Deere manufactured its industrial Model L from 1942 into early 1947, however very few were assembled during World War II years.

RIGHT: 1947 Model B. Deere introduced the styled B in 1938 with a new, stronger, pressed-steel frame. Onboard electric starters allowed Dreyfuss to safely enclose the spinning flywheel.

Iowa, and the new plant opened two years later. Within six months, the Dubuque factory was manufacturing Deere's new Model M tractor.

The other impetus for this pressure to expand facilities and update the product lines came from Michigan again. Henry Ford had returned to tractor manufacture in 1939 Model offering 9N witha three-point hitch. Ford replaced the 9N in 1942 with the Model 2N (for the same pricing reasons that Deere introduced the GM.) The Ford-Ferguson tractor flummoxed Deere's engineers and product specialists.

Ford offered it only as a standard-front low-clearance model. There were no narrow- or wide-front variations. Ford had no plans for tricycle fronts or high crop models either. Yet his small tractor did 50 percent more work per hour than its competitors. It outsold Deere by the same margin.

Ferguson's hydraulic control made Ford's N series the best selling tractor in California. "Ford has stolen the show with his tractor with hydraulic power," Brown noted. Towed implements needed broad headlands to swing around as they reversed direction. Ferguson's integral plow and hydraulics on Ford's tractor let operators maneuver tightly. This increased their yield per acre on California's expensive farmland. California was "the hydraulic state," as Brown learned on a later trip. Visiting dozens of farmers throughout the vast farming regions, he saw countless homemade adaptations on Deere tractors using Ge-Be hydraulics to operate scrapers, wide disk harrows, chisel plows, and even a Case combine. After visiting one 12,000-acre farm, and seeing another that stretched four miles long and one-and-a-half miles wide, Brown understood the labor-saving efficiency that hydraulic power brought to large-scale farming.

Brown continued testing his Model 101 one-row cultivating tractor. He already had fitted it with the next generation of Emil Jirsa's Powr-Trol experimental hydraulics. By July 15, Brown had four prototypes testing. McCormick and another

ABOVE RIGHT: 1949 Model MC. With steady hands the operator works the field in an arrow-straight line. Deere tested and developed these machines for three years over steep hills and rocky terrain to insure reliability. *D&CA*

RIGHT: 1949 Model MC. Deere devised a special course to test and develop the MC, with steep side hills and rocks to throw the crawler tracks off the idler wheels at the bottom. *D&CA*

1950 Model MC.
The little MC pulls a disc through an old-growth vineyard. Deere manufactured the M crawlers from 1949 into 1952. *D&CA*

engineer each spent a day cultivating in small corn alternating between Brown's prototype and an Allis-Chalmers Model B. "They felt the idea of the new tractor is sound, that is, in placing the man to the front. It makes for a much better cultivating job." Brown wrote in his diary that evening. By September 1944, as the war's appetite for men and materiel continued, Brown pushed for a 1946 introduction of his new machine.

When Wiman resigned from Deere to serve in the US Army ordnance department, Deere's board elected George Peek as its new company president. Peek, a Theo Brown fan, led the next quarterly directors meeting. Throughout the session, Peek referred to 101 as the full-vision

tractor or "the Theo Brown tractor." It was, he said, "Deere's first tractor specifically designed for implements."

RETURN TO PEACETIME PRODUCTION

After the war ended, Waterloo engineering reinstated electric start and rubber tires on the Model G and dropped the "M" suffix on its designation Waterloo introduced a variation on the Hi-Crop GH in March 1950, providing the GW wide-front model an additional 14 inches of ground clearance. Waterloo specifically intended the GH for sugar cane growers in the southeast. Deere shipped 73 of these to Louisiana branches for sale and distribution while another 46 went to Florida. The final Model G left Waterloo in February 1953.

Deere continued upgrading and up-rating its venerable Model D, the A, and the B. Unstyled narrow-front and wide-front axle Model As had appeared in 1935, and in 1937, the company had introduced a Hi-Crop version on each front end. The next year Model As got styling and all version continued beneath Henry Dreyfuss's handsome sheet metal. At the same time, Deere introduced the four-speed transmission. In 1941, the six-speed transmissions replaced the four, and Deere increased engine displacements to improve power output.

Deere had introduced the styled version of the D in 1939. It added an optional electric generator, starter, and lighting. Engineering added narrow- and wide-front models and high-clearance versions to the Model B lineup. In 1938, the company replaced the tinwork with styled versions. Where Model B frames formerly had been cast, Waterloo engineering now stamped them out of pressed steel forms. This offered greater rigidity and improved overall quality.

DUBUQUE TRACTORS MOVE THE BENCHMARK

Engineer Willard Nordenson's success in bringing the Models L and LA into production earned him another tractor assignment and a promotion. At Moline, he had devised and developed the prototype S.X./Model M tractors and he led the team that brought them into production. Charley Wiman had been

very pleased with these efforts. In 1946, a year before Deere opened the Dubuque Tractor Works, Wiman named Nordenson to be the new plant's engineering manager.

As Nordenson had done with the L, he mounted the M's two-cylinder engine upright. Its crankshaft ran parallel to the direction of travel. The engine burned gasoline only, running at 1,650 rpm.

Deere & Co. conceived the M as its challenger to Ford's N-series. To compete directly with the Ford, Deere introduced the M first in a standard-front general-purpose utility tractor only. But its power rating created a market that its standard front end didn't completely fill, and at the beginning of 1949, Dubuque introduced the M Tricycle, Model MT. Nordenson's engineers then produced

1950 Model AR. Family resemblance to the Diesel R was intentional when Deere introduced the styled "New Improved AR" in June 1949. It used the same engine block as in the Model D.

an adjustable wide front, a tricycle dual-wheel front, and a single front wheel configuration. What's more, Dubuque engineers expanded the capabilities of Brown's long-awaited hydraulic lift, the Touch-O-Matic. Deere had just introduced this system on the Model M, and the MT offered a duplicate cylinder to provide lift control for left- or right-side implements independently of each other, something the Ford-Ferguson system did not.

Brown's Touch-O-Matic was a fully-integrated hydraulic cylinder system control unit that raised, lowered, or set implements to working depth. Brown and his staff labored for nearly a decade to develop this system, earning Brown more than a dozen new patents along the way.

The tractor engine's crankshaft drove the hydraulic pump. This provided hydraulic functions whether the tractor was moving or not. Operators precisely placed cultivators or plows at the head of a row, and set their working depth before moving forward. A large lever that engineering staffs nicknamed "The

Liquid Brain" stuck out of the transmission case. This lever operated in a quarter-circle arc to control the hydraulics.

Seating position and comfort always concerned Dreyfuss. For the Model M, the industrial designers took their ideas further. Dreyfuss knew that farmers frequently stood while the tractor was moving. Were they stretching tired leg muscles or trying for better visibility in tight maneuvers?

Dreyfuss devised a telescoping steering column with a full 12 inches of travel. Fully forward it cleared room for the operator to stand and lean against the wheel without being pinched. At the longest extension, it allowed the comfortable bent-arm steering style most farmers preferred. Dreyfuss made the seat adjustable forward and aft to accommodate taller or shorter operators who could inflate a small air bag inside the seat to cushion the bouncing or accommodate the tractor's side tilt or list during plowing. Dreyfuss angled the seat forward slightly so operators used their legs to hold them in the seat. This left the thighs unsupported, which encouraged blood flow so legs didn't fall asleep.

BROWN'S 101 FULL VISION TRACTOR

Brown forged ahead with his mid-engine prototype Model 101. In one demonstration after another, he heard "the 101 stole the show." During a dinner following one successful field trial, a number of the observers told him they hoped Deere would put this tractor on the market soon and "be first for once instead of following."

BELOW: 1950 Model M Industrial. Deere manufactured M Industrial tractors from 1949 through mid-1952. The M, as the compact L and LA models had done, mounted its two-cylinder engine vertically.

OPPOSITE: 1950 Model M. Deere engineers did away with a separate frame on the new Model M tractors, instead brazing implement-mounting points onto engine castings. The M also introduced Deere's Quik-Tatch implement connection system.

Brown installed an air-cooled Wisconsin four-cylinder engine into one of the prototypes in late January. In early July, he conceived an industrial and shop version of his creation. He found an affordable four-speed transmission for his 101 and perfected its hydraulic controls. Henry Dreyfuss restyled the sheet metal and redesigned the 101's seat and its position. Wiman (who had returned from the Army in September 1944), Brown, Charley Stone (now head of production for all Deere products), McCormick, and Nordenson conferred about the 101. Nordenson wanted some decision "as to what field this tractor should fill."

Wiman challenged Nordenson: "How would you feel if Allis-Chalmers has a tractor as far along as this?" Brown reported the conversation in his diary, concluding, "Wiman and Stone seem to have more vision about the possibilities than the others." Wiman also sensed the risks.

ENTER THE 102, WITHER THE 101

With the 101 not yet approved for production, Brown had another idea in late March 1945, for a convertible tractor. On this machine, similar to his 101, the operator sat on the engine. However, here the drive wheels straddled the engine. Frame support tubes projected out the front and back. The operator could insert the steerable wheels into either end of the tractor, depending on the job requirements. Front steering worked best for plowing and

harvesting. Cultivating was easier with rear steering. Besides adjustable track, this tractor also provided a variable wheelbase. Brown designed the new machine around Ford-Ferguson hydraulics and lift controls.

As though he sensed that 101 production was unlikely, Brown began pulling back from those prototypes. By late February he was resolved. "Here is what I think we should do with the #101 and tools before we let go entirely. . . . " The seven items on his list amounted to fine tuning the implements. The 65-year-old engineer had developed this tractor as far as he could go. It began to affect his health and by summer, Wiman once again sent him off for another long vacation.

In September, when Brown returned, the 101 was alive again. Wiman had urged

OPPOSITE TOP: 1951 ANH and BNH. Deere's single front wheel helped growers raising vegetables in raised beds. Farmers irrigate these crops by flooding troughs between beds, into which the tractor tires fit exactly.

OPPOSITE BOTTOM: 1952 Model AO. These A Orchard versions were strong tractors, developing 34.1 horsepower on the drawbar. Weighing 4,900 pounds, it was capable of pulling 4,045 pounds.

BELOW: 1951 ANH and BNH. The high ground clearance of these California vegetable tractors allowed plenty of room to operate the machine without damaging tall plants.

ABOVE: 1953 Model GH. Sometimes known as Cotton and Cane tractors, these high-clearance tractors often went directly from the factory to cotton plantations and sugar cane farms from the US Southeast to the Southwest.

RIGHT: 1928 "Low-Down" Concept. Over time, Brown's idea for his "All Visibility" tractor evolved into this low-center-of-gravity machine that offered greater stability and easier access for operators. *GLWPI*

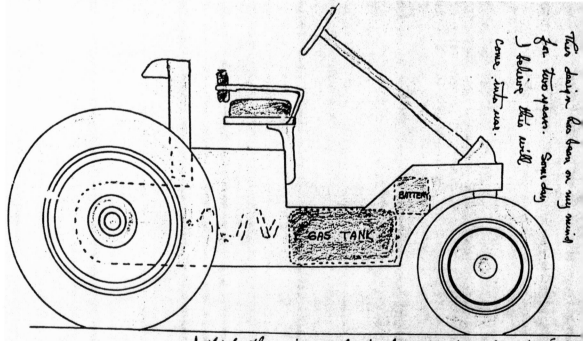

Nordenson to move ahead and he had given it to Bill Cade (Brown's nephew who worked in engineering at Dubuque) and two other engineers. Wiman "said he thought the 101 would be on the market in 1948 and would be needed by that time."

By early October, Nordenson's team had found ways to mount all the hydraulic mechanisms and were encouraged by the tractor's performance. Nordenson "thought the M tractor would be more in demand as a tricycle tractor and the 101 would come in as the big volume four-wheel tractor." Within days Deere started construction of its new Dubuque Tractor building, intending to transfer all tractor manufacture from Moline. Five months later, in mid March 1947, Dubuque completed its first Model M.

Around that time, Brown had one of his prototype 102 models running. He had incorporated Ford-Ferguson hydraulics and hitch elements. He used "remodeled" Ford running gear to operate the cultivator rigs ahead of the front wheels.

During its quarterly meeting in late July 1947, the board voted to increase the Dubuque plant's size by 50 percent (adding some 320,000 square feet) and to enlarge Waterloo by about 30 percent. The goal was to take Waterloo tractor production up from 250 to 325 a day and Dubuque from 50 to 150 machines a day. At that time IHC assembled 600 tractors a day, Ford did 400, and Allis-Chalmer completed 220.

In early November 1947, Wiman's earlier question to Nordenson—what if Allis-Chalmers had a machine similar to Brown's 101?—got an answer:

"Milwaukee; November 4 (AP)—The Allis-Chalmers Manufacturing company will begin production early next year of a new type of farm tractor in a Gadsden, ALA., plant recently leased from the war assets administration. The machine was designed especially for use on family-size farms," the story continued. *"The engine is mounted in the rear and the operator has unobstructed view of the ground working tools mounted on the frame of the tractor ahead of the operator."*

Brown was discouraged but over the next few months he revised his ideas for the Ford-based Model 102 tractor, creating one in late November 1948 with a center pivot, as a kind of early articulated concept.

"This working out new ideas that are somewhat in the future," Brown wrote on April 12, 1949, "is rather discouraging for when our implements and tractors as we build them are in demand, no one relishes the idea of change. Yet I feel strongly that our present line of tractors will be out of date before long. . . ."

1952 Model MC. After the company replaced the Model B with the M, Jesse Lindeman sold his business to Deere and crawler manufacture continued in-house.

Diesel Models

This was the biggest tractor Deere had manufactured so far. It was their most powerful. And it was the one that took the longest time to get right. Deere introduced its diesel-powered Model R at its dealer show in Manitoba, Canada, in June 1949. It wasn't long before the company knew just how "right" they'd gotten it.

In mid-April, before the debut, Deere kicked off the 1949 Nebraska state tractor testing session with Model R serial number 1358. In 57 hours of engine operation, testers reported one single repair: During the warmup run, a glass sediment bowl in the fuel line broke. Engineers replaced it. Testers reported no other repairs or adjustments performed during the duration of the test. They recorded a maximum fuel consumption of 17.63 horsepower-hours per gallon of fuel.

Deere was not the first to offer a diesel. Others had tested them at the Lincoln campus test site. The first was nearly 17 years earlier in June 1932, a Caterpillar Model 65 Diesel. This set a Nebraska test economy record of 13.87 horsepower-hours per gallon. The next year Caterpillar's "Diesel 75" beat

1949 Model R. The diesel engine could only be an electric-start configuration. Its high compression required warming up on gas before shifting fuels to diesel.

1949 Model R. Deere's massive diesel engine set records during University of Nebraska tests. Engineers recorded exceptional fuel economy and rated brake horsepower numbers during 57 hours of operation.

its own previous record with 14.62 horsepower-hours per gallon. Those slipped to second place behind Deere's new R.

Deere's diesel engine development began soon after it enacted trade and marketing agreements with Caterpillar. Customer demand for more power drove the project. Yet the stories of diesel development that slipped out of Waterloo, Moline, and Dubuque factories fueled new rumors of a change from two cylinders to more. It reached a frenzy toward the end of 1936, provoking a strong management response.

Deere sent a letter to its dealers, "Bulletin No. 11", dated December 23, 1936, that emphasized, in brief: "We Are Not Making A Four Cylinder Tractor, Nor Are We Ever Thinking Of Making One."

Developing additional power through the diesel was not simple. The gelatinous nature of the fuel in cold temperatures made starting very difficult under those conditions. Throughout 1936, Waterloo engineering worked on a procedure to start the engine on gasoline and run it until it warmed. Then the operator switched fuel supplies over to diesel. By 1937, engineers were experimenting with a higher-voltage electric starting system using 24 volts. Even that was not enough power to heat the fuel and start the engine in the cold. McCormick's staff tried various combustion chamber shapes in order to produce the required high compression. They needed something like 16-to-1 compared to 4-to-1 for the kerosene engines. By mid-1940, engineers had sufficiently sorted out the combination of problems, and they produced the first of the W.X./Model R prototypes.

Between 1941 and 1945, McCormick's engineers thoroughly tested and evaluated eight R prototypes at test farms in Texas, Minnesota, and Arizona. Waterloo engineering assembled five more during 1945 and another eight in 1947. By this time, McCormick's staff had answered the question of the starting procedure. They devised an opposed two-cylinder gasoline "pony" engine with an electric starter. This resembled a system that Cat's diesels introduced in 1931. Operators ran the pony to warm up the diesel engine block and fuel. Once temperatures reached a certain level, the gasoline engine propelled the starter for the larger diesel main engine.

Waterloo involved Henry Dreyfuss with the R from the start. With the massive cooling required for the twin 5.75x8.0-inch cylinders, the fan tended to suck field debris into the radiator grille mesh. Dreyfuss set the angle of its corrugations so a farmer wearing gloves easily swept them clear. It established a style that customers recognized through the next decade.

Deere manufactured the first diesel R in January 1949 and shipped it to Wolf Point, Montana. Its cold weather starting abilities got an immediate test in Montana's sub-zero winter. But McCormick never got to see his production R drive off the line. He died of a heart attack on September 12, 1948, at age 58. He had been Deere's chief engineer at the Waterloo Tractor Co. for 17 years. Two weeks later, the board appointed Wayne Worthington as head of research at Waterloo, the steppingstone to the chief engineer position.

A TIME OF TRANSITION

Wiman reluctantly approved Brown's retirement on November 1, 1952, but not before exacting a promise that the 74-year-old engineer would serve on the John Deere Advisory Committee. "Of course," Wiman wrote to him, "I may and likely will request that you undertake a particular special assignment from time to time."

1949 Model R. It took a new six-speed transmission to handle the power. At the University of Nebraska, the 7,400-pound tractor pulled 6,644 pounds in first gear.

1949 Model R. The muscular front end of the styled Model D reappeared, refined, on the R. Space between the vertical ridges allowed gloved fingers to pull straw and chaff from the grille.

Deere's Model R ended production on September 17, 1954. Waterloo had assembled something approaching 21,300 tractors. The giant R had developed 45.7 horsepower on the drawbar and 51 horsepower on the belt pulley. The two-cylinder engine, in its many sizes, for all its power, despite its exceptional longevity, had revealed its limitations and showed its age. Protests that the branch sales staffs had issued at the end of 1936, and that the factory had echoed the following spring, no longer reverberated 18 years later.

Waterloo had new tractors ready. For Deere, the production letters sequence stopped at R. Numbers were coming. In the next few years, they grew from tens to hundreds. Soon after that, designations multiplied again, joined by radical changes in engines as well.

Dreyfuss's dramatic sheet metal changes for the diesel Model R established the look that

Deere followed through the 1950s. It hinted at the engineering developments that Waterloo and Dubuque accomplished beneath the skin as well.

The evolution began with a Model 50 to replace the B, the 60 replacing the A, the Model 70 for the G, and the new 40 for the M. It didn't end until the new Model 80 replaced the R. This was a three-year time span that brought countless improvements and advances. The R had provoked engineering innovations and board discussions. The demands for more power were relentless: Should the company reexamine its 25-year devotion to two-cylinder engines?

Electric accessories had proliferated on Deere's tractors. They had starters, lights, and better instrumentation. Waterloo and Dubuque handled this easily with a more-powerful generator and a larger battery. Hydraulics were a different matter. Jirsa and his colleagues at Moline had created the Powr-Trol requiring a pump run off the engine. But operators needed tractors to do more and more work. Visionary implement designers such as Theo Brown, Irishman Harry Ferguson, and IHC's Bert Benjamin steered farmers toward this goal. It took more power just to run the hydraulic pumps that raised and lowered ever-larger attachments.

Deere's two-cylinder engines challenged physics and chemistry. There is a limit to how far a spark can travel across a piston face to ignite a fuel mixture evenly and efficiently. Deere's diesel bought the company a little time. But Deere's board and engineers recognized that diesels gave their two-cylinder engines another decade of life at the most.

Wiman examined other manufacturers as an Army ordnance colonel. When he returned to Deere after the war, he was willing to accept risks. He advocated more cylinders. Nordenson's engineers at Dubuque, anxious for new challenges, encouraged Wiman. They reminded him that Nordenson already had mounted the engines vertically in the Model M. The recumbent two-cylinder engine had lain there, horizontal, longitudinal, unchallenged, since Louis Witry first positioned the side-by-side twin onto the Waterloo Boy chassis almost 40 years earlier.

Board member L. A. "Duke" Rowland argued for patience and caution. Rowland had joined the Board in 1942; six years later it named him manager of the Waterloo Tractor Works. This Englishman knew his own mind and rarely hesitated expressing it. The board had named him a vice president in 1947.

Rowland argued that the time and effort needed to bring Deere's diesel to production had been long and expensive. Although factories might not need the 12 years they spent developing the R for the next project, developing a four- or six-cylinder engine certainly required more than one year.

The Model R had introduced the concept of separate fuel supply for each cylinder (fuel injectors in the case of the diesel engines, carburetors for the others). Mixing the fuel and air so near the intake valves permitted greater precision in blending the two. Horsepower output rose slightly. Nordenson carried this technology over into the new tractors Deere introduced in 1952.

The company launched the Models 50s and 60 first, as row-crop versions. Deere

1949 Model R Diesel.
The work Dreyfuss, his designer Jim Conner, and Waterloo engineers did on the styled tractors is apparent in the strictly businesslike rear end of the R.

immediately offered a Hi-Crop Model 60. Engineers adopted "live" power-take-off for 50 and 60 series models from competitors. Live PTO let the power take-off shaft rotate independently of tractor motion and used an independent clutch to engage it. This kept machinery working at the end of the row where the tractor maneuvered through a turn.

Within the two-cylinder engine, Deere engineers had designed a new combustion chamber referred to as "cyclonic fuel intake." Modifying the intake valve venturi induced a cyclone-like swirl mixing fuel more completely. This increased power and improved economy whether the engine fired on gasoline or distillates. It appeared on the newly offered liquefied propane gas (LPG) versions of the Model 60 in 1954, and in 1955 for the 50.

This better-mixed combustion ran cooler. Still, engineering gave the cooling system a major overhaul. Until the number series began, Deere had not used water pumps; it relied on the thermosyphon system

for cooling. This had served the tractors adequately since the Waterloo Boys. With the new models, Deere pressurized the radiator and pumped the coolant. A thermostat controlled the radiator louvers.

The optional rear exhaust was a Henry Dreyfuss innovation. He maintained that the vertical pipe blew fumes in the operators' faces, blocked their forward vision, and assaulted their hearing. He once posted a note in his New York office that promised a case of Chivas Regal Scotch whisky and a night in the nearby Waldorf Astoria with a famous actress to the first designer or engineer who got the exhaust off Deere's hoods and kept it there. The rear exhaust option routed gases through a series of bends and curves. Eventually the fumes and noise blew out beneath and behind the rear axle. Despite slightly reducing the tractor's ground clearance, Dreyfuss accepted the risk of collapsing the pipe mounted below the belt pulley in exchange for the benefit to operator's lungs and hearing. Deere promoted this for work in and out of low-opening doorways in barns and sheds

In 1953, Deere brought out the Model 70, nearly 20 percent stronger than the Model G that it replaced. The Model 70 and 80 introduced power steering to America's farm tractors. And when the company introduced the 70, the hydraulic Powr-Trol system had one-third more capacity and power than Deere delivered on the Gs. As with the 60 and 50, engineering upgraded the electrical system from six to 12 volts. The Model 70 Diesel again claimed the Nebraska test economy record in 1954 at 50.4 horsepower with 17.74 horsepower-hours per gallon.

BELOW: 1950 Model R. Metal cabs were factory options while dual wheels came from dealers in regions where extra traction was beneficial. Deere introduced the R Diesel in Manitoba, Canada, in June 1949.

OPPOSITE: 1952 Model RI-X. When Deere considered the idea of developing a line of wheeled road construction machinery, it adapted this R Industrial model, adding front mounts for a bulldozer blade.

Models 40, 50, 60, and 70

The Model 40 joined Deere's 1953 tractor line to replace the Model M. This completed the entire lineup. The new 40 series tractors, as well as all the other models in the new number line, were easier to enter or exit. Platform redesign and new fenders allowed room ahead of the rear wheels to climb on or off the tractor. Engineers lowered the center of gravity. Much of Brown's developmental hitch, his implement line, and his hydraulic system for his 102 entered production on the 40, offered as a Standard, Hi-Crop, Tricycle, Wide (two-row utility), Utility, or Crawler, and Deere added a "Special" (using a "V" designation for cotton/sugar cane applications).

The Model 40C improved traction and operator comfort over the MC. Nordenson made four track rollers standard equipment, a step up from the MC's three, and Deere offered five rollers as an option. This improved track and drive gear durability.

1953 Model 40C. An operator shapes a hillside with a Model 40C with a small dozer blade. Deere manufactured the crawlers from 1953 to 1955, and in tests it developed nearly 20 drawbar horsepower.

ABOVE: 1959 Model 530. Deere used a farm in Hanford, California, to develop and test its model HD hay-baler behind a single-front-wheel Model 530N. *D&CA*

RIGHT: 1953 Model 50. Deere manufactured the Model 50 to replace the long-lived Model B. The 50 introduced live power take-off (PTO) and the state-of-the-art Powr-Trol hydraulic implement depth control.

The 80 was the ultimate new model in the series in 1955, retiring the R. Dubuque increased its power output by one-third. The transmission added a sixth forward gear. As with the Model 70, Deere delivered the Model 80 with power steering.

The 40 through 80 series tractors provided power across a broad spectrum. The diesel 80 peaked at 61.8 drawbar horsepower drawbar/67.6 belt pulley horsepower. The LPG Model 70 produced 46.1 drawbar and 52 belt pulley horsepower. Next came the Model 60 LPG with 38.1 and 42.2 horsepower. Then the LPG Model 50 stepped in, peaking at 29.2 and 32.3. The "little" 40C, for its Nebraska tests in early September 1953, pulled 4,515 pounds in low gear, nearly its own weight. It rated 20.1 horsepower on the drawbar and 24.9 on the belt.

THE 20 AND 30 SERIES

For decades the primary function of agricultural tractors has been to pull tools through the ground or across it. Ferguson's three-point hitch had thrown Deere and all its competitors for a loop. It was not merely the success of his device that caused troubles. Ferguson's system had great impact on Ford's N series sales over other tractors. The engineering efficiency and Ferguson's patents on his clever idea for use on Henry Ford's tractors were what plagued the competition. This three-point hitch took a small, standard-front lightweight tractor and endowed it with performance characteristics of a larger, heavier machine by utilizing basic laws of physics. That was another cause of frustration to Ford's competitors. It meant that anything similar to

continued on page 120

1954 Model 40. Deere's adaptation of the three-point hitch made operations like this simple. Here a Model 40 Standard front pulls its single-hank ripper breaking up compacted soil. *D&CA*

TOP: 1954 Model 40C. Farmers–and Deere–had problems with the M series. Fixing those problems and perfecting a replacement led to the 40 series. Model 40 Crawlers appeared as 1954 products.

ABOVE: 1954 Model 40 High Crop. The high-crop tractors got higher clearance with the new 40 series, now offering 32 inches of ground clearance. This was the first Model 40HC, manufactured in late August 1954.

RIGHT: 1954 Model 70HC. The new Model 70 replaced the G series, improving engine output by nearly 20 percent. The Model 70 and its siblings introduced power steering, a necessity on these heavy machines.

ABOVE: **1954 Model 50.** As with previous generations of Deere's wide-front tractors, adjustable axles allowed for work in a vast variety of crop row widths. It was possible to go as wide as 80 inches at the front and 98 at the rear.

BELOW: **1956 Model 60-O.** The "styling" work of Dreyfuss, Jim Conner, and engineers at Deere tapered the long engine cover and sculpted shrouds and fenders over the rear wheels to protect low-hanging branches on fruit and nut trees.

continued from page 117

it, close in appearance or function, was liable for a lawsuit charging patent infringement. Everyone else's version of Ferguson's three-point hitch had to be more complicated.

Deere was no exception. Most everyone played along fairly until the 17-year original life of Ferguson's patent ran out. At that point, a federal court judge refused to renew it. He ruled that Ferguson's three-point hitch was too important to agriculture to remain under patent protection. As the date approached when Ferguson's invention entered the public domain in 1953, each company redesigned its own hitch to take advantage of Ferguson's ingenious linkages and his automatic draft control.

Deere introduced "Custom Powr-Trol." Waterloo and Dubuque engineers had introduced the earlier Powr-Trol system on

the Model R. It did not automatically adjust the implement height or depth to the range of tractor movements over the ground. The operator needed quick reactions to lift or lower the plow when the nose of the tractor dropped or rose up. Deere incorporated "Load-and-Depth Control" into its tractors. This new system offered operators their choice of precisely controlling implement height or depth. They could instead allow the system to adjust working depth automatically in coordination with tractor motion. At this point, Deere had enough major and minor changes to satisfy its product planners and marketing people. To take advantage of the sales potential of these improvements, they designated the new models as the "20" series.

There were other improvements. Brown's Quik-Tatch system rapidly coupled ever-larger implements to Deere's Load-and-Depth Custom Powr-Trol. The operator's seat, already equipped with a low backrest for lumbar support, incorporated a slightly higher seat back, and it added armrests, now available in logo yellow as well as black. For the farmer climbing into the operator's seat on a sunny afternoon in August, the more-reflective yellow absorbed less heat than black vinyl.

Deere introduced models across the complete range. The 320, 420, 520, 620, and 720 appeared in early-to-mid-1956 with the 820 following soon after. The 420s continued the entire line of Model 40 tractors, including the crawler and Hi-Crop version. Deere's sales department had asked Henry Dreyfuss for some new styling.

1956 Model 60-O.
Deere engineers reconfigured engine elements–air intake and exhaust–to keep a narrow profile. Headlights down low had the same intent–to protect the trees.

OPPOSITE: 1956 Model 420HC. By the mid-1950s, Deere was pursuing farmers and distributors worldwide to use, sell, and service their products. This Model 420 High Crop—one of 13—was shipped from the factory to Union of South Africa for its working life.

LEFT: 1958 Model 520. As with its series lineups before this, Deere offered the Model 520 with an adjustable wide-track front end, a row-crop dual front, or this single "tricycle" configuration.

BELOW: 1958 Model 320. Production of the Model 320 series began in late June 1956 with this, the very first one assembled. The comfortably padded yellow Float-ride seat adjusted on an angled track to fit operators.

Dreyfuss knew of other developments working in Moline and Waterloo. The scope of these projects prohibited much investment in current production. Despite this huge new project, Deere's new tractors code named O.X. (4010 series) and O.Y. (3010s) were behind schedule in their overall development. The executive committee of the board reluctantly rescheduled its launch from 1958 to mid-1960. The directors agreed they needed something now to keep customers from going elsewhere for the next shiny new tractor. The board accepted that these changes could be mostly cosmetic.

Jim Conner, the Dreyfuss partner in charge of the John Deere account, recommended painting the engine cover's side panels in yellow. This contrasted with dark green

block letters for the John Deere name and lighter-weight script numbers for the model designation. These appeared on an angle down near the base of the radiator's side panel.

As Deere prepared the 20 series models for introduction, the board learned that not all of 4010 and 3010 work was tardy. Waterloo gasoline engine development specialist John Sandoval had been experimenting with cylinder head and combustion chamber designs. His work yielded so many benefits to performance and fuel economy that Deere introduced some of them on the 520, 620, and 720 models introduced in 1956.

THE TWENTIES BECOME THE THIRTIES

In 1958, Conner trimmed back his vertical side panel. He eliminated the horizontal green bar and gave the engine cover panel a diagonal cut to further expose the engine. The tractor name remained in the same typeface, but he changed the new "30" series numbers to shadow-style, differences bold enough for sales staff to point out.

If any tractors could be called beautiful, they might have been Deere's Models 530 and 630. With these machines, Dreyfuss's staff produced a truly *stylish* tractor. They installed its automobile-style dash panel at the end of a

ABOVE: 1958 Model 620 O. Deere offered farmers a choice of gasoline, all-fuel, or, as on this Orchard model, liquefied propane gas (LPG) to run the engine. LPG had the benefit of lower cost per gallon, and its higher engine compression produced the highest engine output.

LEFT: 1958 Model 830 D. Diesel fuel was proliferating through Deere's entire tractor engine lineup. Near the end of the series, it was powering 820 and 830 models.

LEFT: **1958 Model 430.** In many ways, the 30 series tractors were placeholders. Deere had huge innovations coming but rather than hurry their introductions, these machines offered mostly just appearance changes.

BELOW: **1979 Model 430I.** An industrial tool became a collecting rarity. Deere assembled only 53 of these 430 Industrials with the Holt forklift. An innovative transmission simplified forward-and-reverse direction changes.

flipped-up cowl. They mounted the steering wheel flatter, more like an automobile as well. And they introduced the Human Factor Seat.

Soon after Jim Conner joined Dreyfuss, he worked on airplane seats for Lockheed, collaborating closely with Dr. Janet Travell, a researcher at Harvard's Medical Center. Best known as President John F. Kennedy's back specialist, her less-visible work dealt with human engineering in a manner similar to Dreyfuss's own ideas. She had determined not only how to heal injuries to the lower back, but she also understood their causes. She developed ideas about how to avoid them in the first place.

Placing the tractor operator right over the rear driving wheels was a fundamental

TOP: 1960 Model 840. Starting in 1959, Waterloo manufactured these massive tractor-scraper rigs in-house. This 33-foot-long combination loaded 7.5 cubic yards of dirt in 45 seconds.

ABOVE: 1960 Model 840. An operator tested the Model 840 tractor and Model 400 Elevating Scraper outside of Laredo, Texas. When introduced, the two sold as one item for $14,075. *D&CA*

RIGHT: 1960 Model 8010. When Deere introduced these Model 8010 articulated four-wheel drives in 1959, it was the first full-line manufacturer to produce its own. It weighed 24,860 pounds.

problem. This rear end, often carrying wheel weights for improved traction, provided only tire flex to soften impacts that traveled straight up the operator's spine. The "Float-ride" seat, developed from Travell's recommendations and introduced on Deere's 20 Series, incorporated its own shock absorber on the largest 30 series tractors. She understood that as human height increased, legs and arms were longer. She calculated a 27-degree rising angle figured back from the steering wheel, and Deere's seat rose and fell along an inclined track.

When Deere introduced the 830 in mid-1959, the dealers, sales staff, and customers saw new machines that advanced the art and science of hydraulic draft-controls, engine breathing and combustion, operator comfort, and general ease of operation beyond any competitor. With this series,

more than a decade after Wiman asked Brown to devise a tractor cab, Waterloo introduced a weatherproof steel cab with good visibility and ventilation.

Deere had perfected the diesel engine: at the end of 1959, a review of the top five Nebraska test performers in terms of power and fuel economy, using *any* fuel, counted five Deere tractors. The highest rating went to the Model 720 at 19.97 horsepower-hours per gallon at 56.66 belt horsepower.

By the middle of 1960, Deere had manufactured more than 1,450,000 two-cylinder tractors. The tractor division had achieved breathtaking success. In 40 years, tractors had gone from the single most dreaded topic of boardroom discussion to the corporate profit leader. But few people outside Deere and Dreyfuss understood that the 30 series was just a tease of things to come.

1960 Model 8010. Deere used a GMC 425-cubic-inch six-cylinder supercharged diesel to develop 215 horsepower. Deere considered the 8010s as experiments in future development.

A New Generation of Power

As Deere & Co. approached its 100th anniversary in 1937, Bill Hewitt neared graduation from University of California at Berkeley. He had majored in economics. Following business school studies at Harvard and after World War II, he returned to California and got a job as a territory representative for a Ford tractor distributorship. In 1948, Hewitt married Patrician Wiman, Charley's daughter. He had a job with Deere soon after.

Starting as a San Francisco branch territory manager, he became the branch's assistant manager and then its general manager. He learned the business and the company. Hewitt earned a reputation for sensitivity, grace, even-handedness, intelligence, and fairness. In a company such as Deere, these were virtues that also were part of the company legacy.

1960 Model 4010. Smartly showing off its new appearance, this New Generation 4010 works a field. The tractor developed 84 PTO and almost 72 drawbar horsepower. *D&CA*

In 1954, Wiman was gravely ill. As his health worsened, he created the position of Executive Vice President, setting up his succession. He nominated Hewitt. The board agreed unanimously. Wiman died barely 11 months later, and within two weeks, the board elected Hewitt as Deere's sixth president. He was 40 years old.

Wiman's board had authorized the design work for the New Generation of Power four- and six-cylinder tractors. Before this series, most improvements simply modified existing design. Hewitt knew how the New Generation began.

"There was a meeting," he explained. "Senior engineers, some board members, and the Dreyfuss design people couldn't quite believe that Deere & Company was willing to start fresh, with a 'clean sheet of paper.' One of the engineers asked if that meant they didn't have to carry anything over to the new tractors. The chairman thought and said 'It might be good if the tractors were still green and yellow.'"

CLEAN SHEETS OF PAPER

Waterloo engineers had a design and a scale model of the idea by the time Jim Conner and

1960 Model 3010. This was the first Model 3010 manufactured. Its four-cylinder engine (and six for Model 4010s) was a big surprise. Diesel was the primary fuel choice, although gasoline and LPG were optional.

the Dreyfuss started their work. Waterloo pulled a small group from product engineering and moved them out of the factory to devise the new tractors. Conner recalled the mockup was "a kind of starting point, but we all quickly discarded it," he said.

"It had a narrow-angle V-six engine in it. I don't know if they ever built running engines. It was pretty slim but still the valve covers protruded into the line of sight." But their width and the manufacturing costs, coupled with structural and vibration problems, doomed these six prototypes and the V-configuration.

Waterloo engineering's Merlin Hansen managed Deere's New Design Group that began its work in an empty grocery store before moving in 1955 to the "tin shed." This was Deere's first steel building erected on the site of the new engineering design and test facility southwest of town.

Hansen's assignment came early in 1953. It seemed to present a conundrum: "To give the farmer more tractor for equal or less cost, and to ensure that it would do more work with less effort on the part of the operator." This left plenty of room for interpretation because his assignment included producing a complete power range of machines. With colleagues Wallace DuShane, in charge of overall design, John Townsend, the diesel engineer, and John Sandoval, responsible for gasoline engines, they set targets of 50 and 70 horsepower for two all-new models for the 1958 model year. With Vernon Rugen, they discussed transmissions, agreeing on a broader range of speeds than before in a smaller casing that was easier to shift.

Waterloo's development nomenclature labeled these new machines the O.X. for

the 70 horsepower version and O.Y for the 50 horsepower model. Deere's Name & Numbering Committee gave these models the 4010 and 3010 final designations before introduction. The New Design Group agreed that the 70 horsepower O.X./4010 was its primary development target, configuring it along Model 70-720 dimensions. When Vernon Rugen devised a planetary final drive assembly nearly 11 inches shorter than what Deere used on the 20 series models, Hansen and DuShane reduced the wheelbase and shortened the tractor.

1960 Model 3010. Dreyfuss and designer Conner got involved with the New Generation very early. They made operator platforms fit comfortably so mechanisms worked logically.

continued on page 134

OPPOSITE: 1960 Model 4010. This was the first model 4010 produced. With its six-cylinder engine, it offered farmers 84 horsepower off the PTO. Operators boarded from the front.

TOP: 1960 Model 4010. Dreyfuss and Conner worked with Deere engineers to continue simplifying the "business end" of the tractor, where PTO, mounting links, and hydraulic connections all came together.

ABOVE: 1960 Model 4010. The heart of the New Generation tractor was the 84-horsepower, six-cylinder diesel. Deere used side-mounted radiator grilles, better protecting them from impacts and debris.

continued from page 131

Power output and fuel storage questions involved Waterloo engineers and Dreyfuss designers. Every manufacturer's propane tanks protruded through the sheet metal hoods. Henry Dreyfuss challenged the engineers to find a better location. One of their more radical decisions placed the fuel tank vertically in front of the radiator. This put inches back into the tractor's wheelbase and length, but it improved visibility for cultivating and other front tool bar work.

Conner worked with Deere's Dan Gleeson on chassis, seating, and controls. Concerns over implement hitches led to shifting the operator seat 12 inches forward and lower than on the 720s and 70s, nearer the middle of the tractor. This provided more room for the hitch components. This move also yielded a smoother ride. The forward position also allowed access to the operator's platform from either side of the front of the tractor as well as by stepping over the seat from the rear. The team added steps and handholds on the front side of the platform.

NEW GENERATION, NEW POWER

Merlin Hansen and engine designers Townsend and Sandoval wanted to create engines that shared as many parts as possible. That proved to be an enormous challenge; in the end, the 4010 and 3010 used only the same piston and connecting rod.

Before the engines grew to four and six cylinders, Sandoval and Townsend produced two vertical, inline diesel two-cylinder engines and one gasoline version. Each had

3¾-inch bore and 4⅜-inch stroke. Townsend settled on the Roosa Master rotary diesel fuel injection pump. This is a common system that Ford, Allis-Chalmers, and International Harvester all have used on their diesel engines. The question of cylinder numbers and configurations arose next.

The New Design Group considered an inline four, a 60-degree V-four, a 45-degree V-four cylinder, and another inline four with an overhead camshaft. Development work on the V-engines showed their flaws, and the engineers advanced through four versions of inline four- and six-cylinder engines. Gasoline and LPG engines settled at a four-inch bore and stroke. Townsend continued hunting for a suitable configuration for his diesel, settling on 4⅛-inch bore and 4¾-inch stroke. Sandoval's innovative cylinder head and combustion chamber design already had provided benefits early in the program. Deere's board had chosen to incorporate some of the improvements into production 20-series models starting in 1956. Gas and LPG engines for the 4010

and 3010 got 12-volt electric systems while the diesels needed 24 volts. Sandoval and Townsend designed the engine blocks so oil lines were inside the casting and all fuel lines were external. No fuel leakage could dilute the lubricating oil and damage the crankshaft and main bearings.

The New Design Group planned five different transmissions. These ranged from a basic synchronized gear transmission to a power shift with eight forward speeds and three reverses. As Miller reported, the engineers agreed that new transmissions must transmit 25 percent more power than before, cost no more to manufacture, and fit in a smaller package. Hansen and Rugen considered adding torque converters. They knew that tractor operators do not shift in sequence like truck or car drivers. Instead farmers shift to keep torque consistent. Hansen concluded that "the torque converter was both unnecessary and undesirable for agricultural use."

The scope and scale of development proved larger than engineering had

OPPOSITE TOP:
1960 Model 4010 High Clearance LPG. Deere had a full line of New Generation tractors and implements right away. To better see crops below, the LPG tank sat vertically far forward.

OPPOSITE BOTTOM:
1960 "D-Day." Deere included Industrial wheeled and crawler tractors among the 136 tractors and some 324 implements on display. *D&CA*

BELOW: 1960 "D-Day." On August 29, 1960, some 6,000 Deere dealers and spouses got their first look at the New Generation of Power across 15 acres next to Cotton Bowl Stadium in Dallas. *D&CA*

1960 Model 3010. The New Generation was startlingly new yet in some ways surprisingly subtle. Deere had introduced the flat rear fenders (and dual headlights, like automobiles of the era) on its 1959 Model 530s.

anticipated, and it took longer than Deere's board had hoped. In late 1957, the company postponed introduction to the summer of 1960. This removed the pressure from an engineering staff fearing it might release something not quite ready.

Harold Brock, Deere's new director of research, knew that experience personally. As chief engineer for Ford Motor Company's tractor division, he had watched his engineers slave frantically to complete their Select-O-Speed shift-on-the-fly transmission. The board had imposed an introduction date. The Select-O-Speed offered 10 forward speeds and two reverses. Its problems convinced Brock it was not ready for production. The board pushed back. Brock protested. For his

resistance, the board transferred him over to the truck division and eventually out of Ford. That's when Deere hired him.

Hewitt and Hansen were more patient. They wanted a power shift transmission as anxiously as Ford did. But even months after introducing the Select-O-Speed, Ford's engineers had no clue how to fix transmissions that sometimes failed after just six or eight hours' use. Deere stepped back.

Ed Fletcher and Wilbur Davis both tackled hydraulics for the 4010 and 3010. Deere previously used open center hydraulic systems. These let hydraulic fluid return to reservoirs while trapped oil held the cylinder's piston in place because valves closed. This concept worked well for single-cylinder hydraulic systems. Engineers imagined operators of the new tractors needing separate raise-or-lower capabilities for front and rear or left and right sides. This required two or more cylinders. Fletcher and Davis concluded that the costlier but quicker-responding closed center system was what the 4010 and 3010 needed.

Engineers knew that properly designed hitch links and attachment points meant an implement followed ground contours as the tractor moved. The system they devised blended the best of Brown's work with the greater potential of closed-center hydraulics. Christian Hess had spent years working with Brown on hitch geometry and design, first to get around Ferguson's three-point system, and then to tame and improve it.

With the added weight on the tractor and its capability to lift and carry larger, heavier implements, Fletcher devised a radically new brake system Traditionally, Deere had placed brake drums on differential shafts. These rotated slower and with less torque than the drive axles or the final drives. Fletcher relocated the brake discs within the differential onto the final drive sun pinion and applied a hydraulic boost. This low-torque placement provided much greater stopping ability with Deere's new power brakes.

The New Design Group developed "Styling Objectives" early in the conception phase of this project. Most significantly, these demanded functional styling and a tidy design in which no parts projected beyond the sheet metal. They wanted the rear of the tractor to appear simpler, and the engine frame rails were accessible for mounting implements. Dreyfuss designer Conner explained their approach:

"It was to make a nice-looking tractor where everything had its reason for being," he said. "The curved hood started with the New Generation tractors. Henry proposed and stuck to the idea of making the tractor hood one piece. 'They do it for automobiles,' he said. 'They stamp the entire roof of a car!' This grew into a 'no-visible-joints' recommendation that became a New Design Group policy rule."

The many tire and wheel options included some that put larger wheels on the front so the nose of the tractor was raised. Oversize wheels on the rear made the tractor appear to be going downhill even if stopped. "By having a curved line on top," Conner continued, "you couldn't so easily tell that the tractor was leaning one way or another. And later, with the Sound-Gard cabs, we tapered the cab roof and windows. So there was no longer any vertical or horizontal reference line in the side profile. It didn't look funny or like it was broken when it was tilted down or up in the field!"

1960 Model 4010 Hi-Crop. Prior to releasing the tractor to customers, Deere extensively tested the Hi-Crops in Culiacan, Mexico. Here an operator works with border discs to bank soil against tomato plants. *D&CA*

JOHN DEERE

presents

the New Generation of Power

35 H.P. Ten-Ten series
Row-Crop, Crawler, Utility Models

45 H.P. Twen
Row-Crop, Row-Crop Utility

A Size for every Farm... A Type for every Jo

Such detail seems inconsequential until compared with the competition. Dreyfuss understood that most tractor purchases involved the whole family. Farmers might consider Deere's tractors were the mechanical equal to IHC's or Ford's. However, if Deere's looked better, less "funny," in the field pulling plows or running home down the road for dinner, that factor might cinch the sale.

Sales certainly was the goal on D-Day, August 29, 1960.

That Monday some 100 buses from 75 cities delivered nearly 6,000 John Deere dealers and spouses to Dallas, Texas. The next day, when Bill Hewitt opened the back door of the convention center and the faithful spilled out, they saw $2 million worth of Deere's New Generation of Power Model 4010 and 3010 tractors and accessories. Waterloo Tractor Works displayed 136 tractors, and 324 supplementary implements, tools, wagons, and combines came from Deere's other plants. They spread across a 15-acre parking lot next to the Cotton Bowl stadium.

The New Generation of Power reached farmers everywhere by 1961. Deere introduced tractors ranging from the six-cylinder diesel 4010 with 80 horsepower at introduction through the 55-horsepower 3010, down to the 35-horsepower gas or diesel four-cylinder Model 1010. The production 4010 developed the power of a Model 830 yet it almost fit inside the dimensions of the smaller 730. The 3010's introductory price was in line with Deere's suggested retail price for its earlier 70 and 720 models.

By 1962, Deere had supplemented the lineup with the 5010, (called O.Z. during development.) This was a 100-horsepower,

six-cylinder, two wheel-drive tractor rated to pull seven plows. In addition, Waterloo offered an orchard and grove version of the small 1010 and of the 55-horsepower four-cylinder Model 3010. But once again, neither Waterloo nor Dreyfuss chose to rest on their newly earned laurels.

One year later, Deere organized Flight Sixty-Three, bringing dealers to Waterloo. Sales had slackened slightly and Deere management gambled that giving distributors a tour of the factory and showing them new tractor models on the production line might spark their initiative. If that wasn't enough, the board hoped introducing the new models might energize branch sales staffs. Waterloo's engineers increased the 3020's output from 59 to 65 horsepower out of the four-cylinder engine. It raised the already respectable 4020 six-cylinder from 84 to 91 horsepower. Waterloo provided an optional Power Differential Lock.

OPPOSITE: Deere redesigned and re-engineered its tractors to create the New Generation, and it updated parts counters and literature. The artwork behind the counterman teases the eye. *D&CA*

BELOW: 1964 Model 1010 Orchard and Grove. Deere based its new 1010 series orchard tractors on its versatile 1010 Utility models, adding rear fenders and using the step as a branch deflector.

NO LONGER RUNNER UP

The New Generation lineup fulfilled Hewitt's longest-held ambition. Deere had sold 23 percent of the farm tractors in the United States in 1959, the last year of its two-cylinder models. By January 1964, the company's market share had reached 34 percent. Deere had become the largest farm equipment manufacturer in the world, passing long-time rival International Harvester.

Barely six months later, Deere opened its new Administrative Center in June 1964. Finnish architect Eero Saarinen had designed the striking building just before he died in 1961. His associates completed his plans to use unpainted steel girders as exterior structural and design elements. Saarinen had selected the steel material, called Cor-Ten, because it would rust quickly and its deep red color would blend into the rolling site.

UPGRADES AND IMPROVEMENTS BECAME STANDARD PROCEDURE

In 1965, Deere upgraded the 5010 to 5020 nomenclature as Waterloo raised output from 121 to 133 horsepower. In 1969, it increased output to 141. Deere offered the 4020 and 3020 with optional hydraulic-powered front-wheel drive, called Front Wheel Assist, or FWA. This was not a true four-wheel drive, but it significantly improved traction. With modifications inside the engine, engineers increased 4020 output to 96 horsepower and to 71 for the four-cylinder 3020. Waterloo introduced turbochargers on the diesel 4020 models in 1968, renumbered as the 4520 with 122 horsepower. The 4320 arrived in 1971 with 115 horsepower, and the 4620 superseded the 4520 with 135 horsepower. These looked virtually identical to the first New Generation tractors except that Waterloo fitted the FWA models with smaller front tires that were chevron-patterned versions of the rears. Engineering substantially reinforced the front axle housings in front of and around the differential casing.

In 1968, Deere introduced replacements for its big articulated four-wheel drive 8010 and 8020 models. These came from Wagner Tractor Co. of Portland, Oregon. Wagner's primary market, the Pacific Northwest logging industry, was in a recession and the company considered laying off employees. Waterloo engineering had encountered delays in developing its replacement four-wheel drives. Moline contracted with Wagner, taking all Wagner's agricultural (WA) production of its two largest machines, the 178-horsepower WA-14 and the 220-horsepower WA-17

1964 Model 4010 Industrial. Deere no longer used Holt or other outside forklift manufacturers but instead produced everything in house, from front weights to frames to operator overhead protection.

for 1969 and 1970. Wagner painted these giants in Deere Green and Yellow.

By 1970, Deere's 7020 was ready for release, and the company ended its relationship with Wagner. Waterloo's new machine was less powerful with 145 turbocharged horsepower, but it was more technologically advanced. Deere introduced the 7020 in 1971; it remained in production through 1975, when Deere intercooled the turbocharged fuel mix. This brought another technology and a substantial power boost to farm fields. Called the 7520, this high-tech model arrived with 175 horsepower in 1972.

1965 Model 4020 LPG High Clearance. The new 4020 not only provided a bit more power but also introduced the Power Shift transmission offering eight forward gears shifting on the fly.

10

Generation II Tractors and the 21st Century

Farmers understood the logic behind a rollover protection system typically known as ROPS. Yet they steadfastly believed they would never need it. Operator safety had concerned Deere since Theo Brown's early efforts to cover rear PTO shafts. In mid-1959, engineer Roy Harrington and product development manager Charles Morrison fabricated Deere's first ROPS, with seat belts, on a Model 4010 prototype. New Generation chief engineer Merlin Hansen understood that if Deere disguised safety improvements (that no farmer ever asked for) as a sunshade (that many had requested), the customer might even pay for it. Deere introduced its Roll-Gard structure in 1966. Even with its sunshade-style roof, it met few takers. Deere took brash steps to make the safety device more appealing.

"To install a proper roll bar would add five hundred or eight hundred dollars to the cost of

1969 Model 4520. Deere's Model 4520 was its first to use a factory-mounted turbocharger. The 6.6-liter inline six cylinder diesel produced 120 horsepower at the PTO.

1966 Model 4020 with Roll-Gard. Rollover protection was something every farmer knew the other guy needed. Deere developed this system to introduce with New Generation tractors.

the tractor," Bill Hewitt explained. "But instead of paying eight hundred dollars more for a Deere tractor, they would buy another tractor from someone else.

"We decided to give all of our engineering specs to each of our competitors," he continued. "They wouldn't have to go through the process of testing and developing [their own] on the agreement that they would make roll bars standard equipment and we would make them standard equipment. Because we couldn't sell the things in any quantity to justify our investment in them, we gave away our engineering. Everybody then adopted roll

bars. Then the cabs came in with glass and air conditioning. The roll bar was incorporated and we couldn't make them fast enough."

Jim Conner picked up the story from the Dreyfuss perspective. Their assignment with Generation II tractors was to integrate the Roll-Gard and the cab so it was strong enough to survive a rollover.

"That was the beginning of the Sound-Gard cab," he said. "We decided to do the best cab possible with all the safety features, all the human factors features that a cab could have: roll-over protection, sound insulation, heat insulation, operator insulation from vibration and dust, air conditioning, pressurization so the air goes out, does not get sucked in, improved vision, easy entry and exit, flat platform, the second generation of the Travell-designed seat." It was quiet enough that Deere's cassette stereo radio became a popular option.

Once Waterloo put the Sound-Gard in production, Deere offered it on the four Generation II "30-series" models introduced in mid-August 1972. The updated cab inspired a new name for Deere's tractor engineering as well, the Sound-Idea line. The Generation II hood was much lower at the front of the tractor than New Generation tractors had been. "To let it slope downward gave you better visibility over the nose," Conner explained. What's more, "it gave you more space under the back of the hood to get 'stuff' in, because there is always more 'stuff' that needs to be in there."

The new Model 4030 offered 80 horsepower, the 4230 offered 100, the Model 4430 produced 125, and the 4630 rated 150 horsepower at the PTO. Deere made Synchro-Range transmissions standard equipment, connected by the Perma-Clutch, the innovative hydraulic wet-plate clutch. For the farmer needing even more power, the articulated four-wheel drives provided 175 horsepower from the 8430 and 225 from the 8630. By the mid-1960s, when work began on the 30 series, Waterloo planned to limit production to diesel engines exclusively for these machines. At the end of 1973, Waterloo ended gas engine

ABOVE: 1966 Model 4020 with Roll-Gard. Deere tried to soften the safety lecture by labeling the structure as a sunshade, something many farmers had requested. Still, few paid the money.

RIGHT: 1966 Model 4020 with Roll-Gard. Ultimately Deere Chairman Bill Hewitt gave Deere's ROPS research and design to all competitors if they would also introduce it as standard equipment.

production. Demand for Generation II tractors forced Waterloo to expand its production facilities. It built an additional engine manufacturing plant as well.

Deere upgraded its entire tractor line in mid-1975 with the introduction of its 40-series. This range began with the 40-horsepower three-cylinder diesel Model 2040 manufactured at Deere's Mannheim, Germany, plant. Mannheim also provided the three-cylinder 50-horsepower 2240 and the six-cylinder 80-horsepower 2860, which appeared in early 1976. The midrange four-cylinder 60-horsepower Model 2440 and 70-horsepower 2640 came from Dubuque.

Deere named its bigger 40 series tractors Iron Workhorses. These powerhouses appeared as 1978 models. The smallest, the Model 4040, used an all-new 404-cubic

inch (ci) diesel that developed 90 horsepower on the PTO. The rest of the series used the 466-ci block. Deere's Model 4240 produced 110 horsepower, the turbocharged 4440 put out 130, while Deere rated the big 4640 with turbocharger and intercooler at 155 horsepower. The biggest of these was Waterloo's 4840 with 180 horsepower and the Power Shift transmission as standard equipment. For big power, nothing beat the four-wheel-drive 180-horsepower 8440 and the 8640 with 228 horsepower. These articulated models offered optional front and rear hydraulic differential locks for really tough traction conditions and as many as four remote hydraulic cylinder outlets. Deere's new single-lever-operation Quad-Range transmission got so much torque to the ground that operators discovered their tires limited traction. Dual- and triple-wheel configurations appeared on large farms throughout the Midwest and west.

New styling on the hoods marked the obvious difference between the early 40-series and the "New Profiles in Performance" midrange models introduced for 1980. The hoods angled more noticeably up to the dash that incorporated an electronic instrument panel. Model designations changed only with the 80-horsepower Mannheim six; it became the 2940. Mannheim engineering offered each of its tractors with optional mechanical front wheel drive (MFWD). They moved the front differential to the side and mounted the tie rod ahead of the front axle to reduce damage to young plants.

Starting in the late 1960s, Deere's experimental researchers investigated alternative fuels and power sources. These ranged from gas turbines to rotary engines.

OPPOSITE TOP: 1969 Model 820 Vineyard. Deere repositioned the front axle farther back and narrowed the track on these Vineyard models for French markets. In this configuration the tractors fit between vineyard rows.

OPPOSITE BOTTOM: 1969 Model 4520. The Model 4520 was Deere's first to incorporate a turbocharger. With 107 drawbar horsepower, it made pulling this two-way four-bottom plow easy work. *D&CA*

TOP: 1970 Model 7020. By 1970, Deere engineers had improved the 6.6-liter diesel to the point where, for these Model 7020 articulated tractors, the engine developed 146.2 horsepower on the PTO.

ABOVE: 1970 Model 7020. Deere assembled 2,586 of these machines. By 1975, the last year of production, they sold for $22,000 with the standard cab.

With each, the results indicated that diesel engines remained the most economical and efficient. Engineering steadily increased Deere engines' output and improved their economy. In 1982, Deere introduced its 50-series. These ranged upward from the Mannheim-built 45-horsepower three-cylinder 2150 and 50-horsepower 2255 orchard and vineyard models to the 300 PTO-horsepower articulated four-wheel drive Model 8850. This used a new turbocharged, intercooled V-8 designed and manufactured at the new Waterloo engine plant. Deere considered 100-190 horsepower its midrange, and for the five new models in this series, Deere introduced a new 15-speed power shift transmission that boasted a 7 percent improvement in fuel economy.

Deere offered its innovative Castor/Action MFWD on all the models starting with the

ABOVE: **1971 Model 1520 Orchard.** Deere designed this low-slung orchard model for European markets. American farmers preferred its size to the smaller 1020 and the much larger 2020.

RIGHT: **1971 Model 2520H.** Deere assembled just eight of these high-clearance models with gas engines and another 98 with 3.6-liter four-cylinder diesels, producing 61.3 horsepower off the PTO.

1971 Model 4020 with Front Wheel Assist. The 4020 series took 4010 introductory horsepower from 84 to 91. Front Wheel Assist, a kind of four-wheel drive, was a very effective way to use all that power.

Mannheim four-cylinder 55-horsepower 2350, up through Waterloo's six-cylinder, intercooled, 190-horsepower 4850. This system added 13 degrees of front wheel caster, or rake, to facilitate extremely tight turning. By laying the front tires over in a turn, tractors carved a smaller circle, something otherwise impossible with four-wheel-drive tractors whose drive gearing binds up in acute angles.

Deere returned to specialty crop manufacture with the low-profile 2750 for orchards, the high-clearance 2750 "Mudder"

for California vegetable farmers working with bedded crops, and a series of wide-tread models on the 2350, 2550, 2750, and 2950 chassis for other vegetable and tobacco growers. New Sound-Gard bodies were even quieter than ever, measuring 73.5 dB(A) under full load.

The 55-series arrived in time for Deere's 150th company birthday in 1987. The Mannheim line filled in mid-power-range models, stretching from the three-cylinder, 45-horsepower 2155 through the

LEFT: 1971 Model 4020 with Front Wheel Assist. The Model 4620s added an intercooler to the 4520's turbo, making the fuel charge denser. PTO horsepower increased to almost 136.

BELOW: 1972 Model 3020H. Waterloo's plant concluded Model 3020 High Crop production with this tractor although 4020 manufacturing continued into 1972.

turbocharged, three-cylinder 2355N with 55 horsepower, specially configured for orchards and vineyards. Deere renamed the 2755 "Mudder" and its companion 2955 as high-clearance models. An optional 96-inch wheel track turned these models into California vegetable specialists. Waterloo added its own vineyard model, the 2855N, at 80 horsepower. Many California wineries ordered these with MFWD to pull one- and two-ton grape wagons around hilly vineyards during harvest.

In January 1989, Deere sent its dealers to Palm Springs, California, where it introduced the big 55-series machines. These appeared as a 105-horsepower Model 4055, the 120-horsepower 4255 offered as a high crop, as well as a standard or MFWD

model. The lineup stopped at Deere's first 200-horsepower row-crop model, the 4955. Each of these used Waterloo's 7.6-liter inline six-cylinder diesel with redesigned intake and exhaust valves and ports and new seven-hole fuel injection nozzles. All models but the 4955 used the 16-speed Quad-Range transmission and Deere's Perma-Clutch. The big tractor came standard with the 15-speed Power Shift.

Deere's biggest tractors, the new 60-series, premiered in Denver prior to the Palm Springs winter introduction. Deere offered a 200-horsepower, articulated four-wheel drive Model 8560 as an alternative to its row-crop 4955. Articulated power ran up as high as a monstrous 14-liter Cummins turbo diesel for the 322 PTO-horsepower Model 8960. Waterloo engineers moved the air intake stack and the exhaust and switched to a one-piece upper windshield, changes that improved forward visibility. The 60-series designation filtered down to Waterloo's row-crop version beginning in 1991. Models designated as 4560, 4760, and 4960 replaced the 4555, 4755, and 4955. These represented innovation in an atypical fashion. The new models offered no upgrade or improvement in horsepower output. But the other changes, updates, and modifications convinced sales and marketing staffs that these were, indeed, new models.

For decades Deere engineers and Dreyfuss designers had labored to eliminate the blind spot that the air intake and exhaust pipe caused tractor operators. With the new 60 series, Waterloo mounted the air intake under the hood and gave it easy access through an opening side panel to clean it or change the element. It relocated the exhaust pipe to the front right pillar of the Sound-Gard cab.

Engineers and designers reconfigured the Sound-Gard entry with a larger platform, two handrails, and steps that adjusted to accommodate frame-mounted equipment or crops. Optional exterior lighting not only illuminated the steps but also fully flooded the entire 360-degree circle around the tractor, making night crop transfer from combines onto waiting trucks easier and safer.

Deere brought back Brown's original rear axle width adjustment system, though it was greatly updated. Still, all an operator needed to do was jack up the rear wheel. After releasing the locking bolts and securing the jacking screw, the operator put the tractor in gear and watched as the wheels racked from a 60-inch tread out to 134 inches or back in.

COMFORTGARD AND THE ALL NEW BREED OF POWER

ComfortGard cabs replaced Sound-Gards in 1992. The company's "All New Breed of Power" 6000 and 7000 series tractors brought more power and greater sophistication to mechanized farming. New inline six-cylinder turbocharged diesels delivered as much as 38 percent more torque from the 66-horsepower Model 6200 up to the 145-horsepower 7800.

The 70-series articulated models delivered even more. With at least 250 horsepower on hand in the 8570 and more than 400 in the 8970s, Waterloo engineers made good use of new electronics. The Electric Engine Control

system electronically adjusted valve timing to add as much as 25 more horsepower beyond the tractor's advertised output. This could occur only when the engine speed dropped below 1900 rpm and sensors detected lugging. As engine speed reached 2100 rpm, timing would return to normal and power dropped to rated output. Deere called these models "Power Plus." It heralded its 1994 powerhouse 8000 series as "21st Century Technology Today."

In the late 1980s, Deere acquired land and designed and constructed a large tractor-manufacturing plant at Augusta, Georgia. Its first products emerged in 1991 and 1992 as Generation 3 machines, starting the Thousand Series with the 5200 with 40 horsepower, the 50-horsepower 5300, and the turbocharged 5400 model rated at 60 horsepower. Engineers moved the seat and the console-mounted controls and instruments even farther forward by relocating the fuel tank behind the seat. Designers fitted steps on both sides of the steering wheel.

Waterloo and Mannheim continued to update models for Europe and the US through the mid-1990s, and they changed the entire lineup to the Thousand Ten Series for 1998 and the Thousand Twenty Series for 2000 and 2001. These included models from an 18-horsepower 4010 in 1996 to the 450-horsepower Model 9520 in 2003.

DEERE ON TRACKS

Never short on innovation, Deere & Co. startled the agriculture and construction worlds in late 1996 when it introduced a rubber tracked model as part of its big tractor lineup for sale in

1974 Model 4230 LP. The "LP" designation is "low profile," renaming the Orchard models for the Generation II series. Deere assembled 90 of these LPs and some Hi-Crops.

RIGHT: 1975 Model 2130 AS. This Mannheim-built tractor with Mechanical Front Wheel Assist uses a Sekura cab from Denmark. Front tires are mounted backward for traction in reverse.

BELOW: 1975 Model 8430. These big 23,000-pound Generation II four-wheel drives developed 175 horsepower and could pull 22,500 pounds! The Sound-Gard cab was standard.

1997. The rubber-belted crawlers got a "T" suffix following their model number. Deere proceeded cautiously with development of its new models. It announced its 9000 series four-wheel drives in 1996. After delaying introduction, the 9000T series crawlers appeared for sale in 1999. Deere produced four-wheel drive and belted-track versions of each model as well, from the 160-horsepower 8100 through the 225-horsepower 8400.

Deere's entire lineup continued up the counting progression in 2001 when all its lines reached the new Thousand Twenty Series. The 5020 range of utility tractors started at the 5220 with 45 horsepower at the PTO up through the 5520 delivering 75 horsepower, and they were available in the spring. Deere introduced its dealers to the Advantage Series, including the 6403

85-horsepower four-cylinder and the 6603 with a 95-horsepower six-cylinder diesel, at its national show in Albuquerque, New Mexico, in August. Dealers also saw new 6020 Series models starting with the 6120 with 80 horsepower up through the 6420 with 110 horsepower. The powerhouses of the lineup also arrived at Albuquerque, ranging from the 8120/T with 170 horsepower up through the 8520/T with 255 horsepower and the bigger-still 9020s, from the 9120/T with 280 horsepower up to the 9520/T with 450 horsepower. Stronger engines, hydraulics, and hitches, better brakes, tighter maneuverability, virtually negligible soil compaction with its tracked models, and unsurpassed operator comfort and sophistication from the Independent Link Suspension to the Active Seat characterized Deere's latest tractors.

1976 Model 2130 AS. Deere assembled these machines in its Mannheim plant using engines from its factory in Saran, France. While popular in Europe and Canada, few reached America.

1979 Model 4240. The 40 series was the row-crop model in Deere's Iron Horse series launched in 1978. Engineers added iron and more horsepower for field performance and durability.

The proliferation of tractors continued just months later when in early 2002, Deere introduced a wide range of nine tractors in the 4000 Ten Series, divided into three chassis sizes: small, mid, and large. These started with the small-chassis 4010 with 18½ horsepower and culminated with the large-chassis model 4710 delivering 48 horsepower. The small, mid, and large chassis designations then carried on into the 2000, 3000, and 4000 product lines. But the company wasn't finished. The 6015 Series offered four new tractors, starting with the 6215 (with 70 horsepower on the PTO) up to the 6715 with 105 horsepower.

The 7020 Series arrived with output ranging from 95 horsepower in the 7220 up to 125 horsepower for the 7520. A new engine line offered in the same model range, the 6.8-liter PowerTech series, delivered from 110-150 horsepower.

Realizing a long-held dream to better match engine performance to ground travel, Deere introduced its Infinitely Variable Transmission (IVT) in early 2003. It initially offered this "shift-less" transmission as an option in the 7710 and 7810 models, offering speeds from 0.03 mph to 26 mph or *any* speed in between. It became standard equipment on the 7920 model. The Infinitely Variable Transmission had special programs to maximize engine effectiveness for PTO use, in heavy draft and tilling applications, and for light tilling or transport needs.

OUTPUT TOPS FIVE HUNDRED HORSEPOWER

Deere raised the bar again in August 2003 when, at its dealer show in Columbus, Ohio, it unveiled the 500-horsepower 9620. Further down the scale, the 5003 Series emerged from Deere's Augusta, Georgia, assembly plant; the three models ranged from the 5103 with 43 horsepower up to the 5303 with 55 horsepower. Farmers, tractor enthusiasts, and historians began to feel the need for a computer database to keep track of Deere's ever-expanding lineup when, for 2004, the company introduced the 5015 Series, offering

some special-purposes machines, with vineyard (V), orchard (O), and Hi-Crop versions using turbocharged three- and four-cylinder Tier II engines. Deere improved its 4000 Series in the replacement 4020 range, offering the 4120 with 36 horsepower (at the PTO) up through a 50-horsepower 4720. It did the same with the 7010 models, now designated 7020 and starting with the 140 PTO horsepower 7720 and ending with a new 7920 with 170 horsepower.

For 2006, Deere offered the 8030 series as tracked or wheeled variations, starting with the 8130 delivering 180 PTO horsepower up to the 8530 with 277 horsepower. The tracked 8430/T with PowerShift transmission, which Deere rated at 255 PTO horsepower, took home honors from its Nebraska test over three weeks in October 2006 when it became the most fuel-efficient tractor the university had tested to that point. It sipped a frugal 14.4 gallons per hour during PTO tests—in which

1979 Model 8640.
Introduced as the top of the Iron Horse line, the 8640 delivered 203 drawbar horsepower, easily pulling this disc-ripper through corn stubble.

it developed a maximum of 298.2 horsepower. For the test, it averaged 258.54 PTO horsepower. It achieved 18.65 horsepower-hours/gallon, an 8.8 percent increase over the previous record holder, its predecessor, the 8520. In 2007, Deere introduced its Premium designation, beginning with the 6030 models and adding 7030s, topping out with the 7430 at 140 horsepower.

Model designations changed once again in mid-2008 as Deere announced its adoption of a Worldwide Numbering Scheme. The first digit of a model number designated its size while the next three listed engine output. A final letter described specification level. For example, the new 5055D was a five series 55-horsepower base model. D tractors were only two-wheel drive; Series E Limiteds (the next level up)

MFWD; M series were mid-specification tractors; and R models were the highest specification, also known as the Premium models. One final letter, T, designated tracked machines. Deere stated engine horsepower from then on in metric measurement.

COVERING THE GLOBE

In 1924, General Motors chairman Alfred Sloan told shareholders that GM's goal was to make "a car for every purse and purpose." As the twentieth century ended with Deere & Co. holding the worldwide lead in tractor manufacture, it seemed to adopt that philosophy as well. For decades Deere has manufactured machines throughout the world, and for more than a century

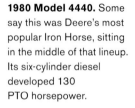

1980 Model 4440. Some say this was Deere's most popular Iron Horse, sitting in the middle of that lineup. Its six-cylinder diesel developed 130 PTO horsepower.

ABOVE: 1981 Model 4640. The 4600 series immediately followed the 4400s. PTO output rose to 155 horsepower, useful as it emptied its hopper alongside the 8820 combine.

LEFT: 1985 Model 4450 Caster Action MFWD. Deere's Caster Action allows the tractor's front wheels to turn as much as 52 degrees, during which they incline slightly into the soil for better traction.

the company has acquired complementary manufacturers to further Deere's interests and expand and improve its product lines. As early as 1956, the company acquired a Mexican company manufacturing small tractors. In the same year it took a majority interest in Heinrich Lanz, a German tractor producer that also assembled machines in Spain. In 1988, Deere entered a joint venture with Hitachi Construction Machinery to produce excavators in the United States. A decade later, the relationship expanded to include manufacturing logging machines based on Hitachi excavators. This venture assembles machines in Canada and Brazil as well as the United States. In 2000, Deere opened a new tractor assembly facility in Sanaswadi, Pune, India, and in 2006, it launched Tianjin Works in China to manufacture transmissions. One year later, the company acquired a tractor manufacturer in Ningbo, China.

To make all these relationships beneficial, Deere also established sales branches throughout the world, beginning in Spain in 1964 and the United Kingdom and Ireland in 1966. John Deere Export attended to customers in Africa, the Middle East, and the European Community starting in 1967 until the company opened specific offices in

ABOVE: **1975 Model 2755.** Deere manufactured 2755s in Mannheim, Germany, or Saltillo, Mexico. Some call these High Clearance models California Mudders for their work in deeply bedded vegetable fields.

LEFT: **1990 Model 4555.** Mechanical front-wheel drive and 155 PTO horsepower enable this tractor to mow and rake corn stubble in a single operation.

OPPOSITE: This 2140 sprayed pesticide in the long evening sunlight of Skane, Sweden, in mid-May, 2017. Deere produced these 82-horsepower four-wheel drives in Mannheim from 1980 through 1987. *Shutterstock*

ABOVE: Deere manufactured the Model 2850 from 1986 to 1994 in Rosario, Argentina, and Mannheim, Germany. This German-born 86-horsepower utility tractor was displayed at the 2014 annual agricultural equipment demonstrations near Salo, Finland. *Shutterstock*

LEFT: A farmer on a small farm outside Haute-Savoie, France, parked his Model 3350 alongside his corn dryer for future animal feed. Deere produced these 100—engine horsepower utility tractors in Argentina and Germany from 1986 through 1993.

OPPOSITE: **1994 Model 770.** Yanmar, a Japanese manufacturer, produced these compact utility tractors for Deere from 1989 to 1998. The three-cylinder diesel engine developed 20 PTO horsepower.

France, Germany, and Italy in 1968. During the economically troubled 1980s, European manufacturers consolidated efforts or quit during the 1990s, and since the fall of the Berlin Wall in 1989, Deere has reached into Central and Eastern Europe and established the John Deere Russia Branch and later a marketing office in the Ukraine. In 2010, Deere opened a manufacturing and warehouse facility in Domodedovo, Russia, south of Moscow, serving not only Russia but also Far East customers. In a complex procedure, it assembled its first tractors in Russia beginning in April 2013. As David Larson, general director John Deere Russia, explained, these are machines manufactured in Waterloo,

tested, and then disassembled for shipment to Domodedovo, where workers reassemble them, test them again, and distribute them throughout Russia and the region. "This local assembly merits a lower customs duty than for importing a finished tractor," he said.

Deere celebrated its 175th anniversary in 2012. Kicking things off, Deere chairman and CEO Samuel Allen rang the closing bell at the New York Stock Exchange on Friday, June 15. The company staged two days of demonstrations by historic re-enactors August 2 and 3 at the John Deere Historic Site in Clinton, Illinois. Visitors saw basket weaving, broom making, blacksmithing, and gunsmithing.

But that was technology from nearly two centuries back. As Deere entered the 2010s, it drastically expanded computer use beyond engine and transmission management to encompass field and task monitoring and management. The Gen 4 Command Center, introduced in late 2013, provided the latest in touch-screen capabilities to manage and adjust tractor settings (from audio through lighting to suspension to transmission), applications (from AutoTrac guidance through implement and machine profiles to a machine and a work monitor), and systems adjustments that let operators change or reconfigure the display screen elements. Deere introduced the devices with a 10-inch screen (the 4600) or a 7-inch version (4100.) Within a short while, improvements and updates to the system allowed

---

ABOVE: 1995 Model 8100 MFWD. The Caster Action is visible here as this 8100 MFWD tractor maneuvers during corn harvest. The six-cylinder diesel engine developed 160 PTO horsepower.

RIGHT: The 235-horsepower Model 8435RT participated in demonstrations during the annual August Agricultural Harvesting and Cultivating Show at Salo, Finland. *Shutterstock*

farmers to upload data in real time to home computers and a massive data and information and resource bank at Deere where, at user request, company consultants analyzed the data and returned observations and suggestions to improve work quality and efficiency. The "Premium Activation" was available from the beginning of 2016.

In keeping with the new designations, 6E Series tractors—from 6105E through 6135E—reached dealers in 2016, as did the 6230R and 6250R and the big 8400R and massive 9R models, including 9570RX and 9620RX machines. Right on time, 5R series tractor Models 5115R and 5125R arrived at dealers in early 2017. The progression continued through 2017 as the company headed to its historic landmark in 2018.

Deere has persevered. While not all its products have been perfect, the company has scored more often than it has missed.

Nearly all the companies it once competed against are gone, merged into acronyms or hyphenates, or submerged into other corporate identities.

Deere & Co. commemorated John Deere's 200th birthday in February 2004. It celebrates 100 years manufacturing tractors in 2018. For a company with a heritage as rich as this one, it is natural to sit back and reflect on past accomplishments, on nineteenth and twentieth century advances and successes.

What should comfort Deere's tractor owners and farmers is a certain knowledge. Somewhere in an office in twenty-first century Moline or Augusta, in New York City, or Mannheim, or Waterloo, engineers and designers are hard at work on computers, on paper, in their heads, developing twenty-second century technology for Deere's tractors next year and years to come.

OPPOSITE: 1998 Model 9100. These were big four-wheel drives, weighing 28,000 pounds and providing more than 214 PTO horsepower. In low gear these pulled nearly 34,000 pounds.

BELOW: 2001 Model 9300T. Deere's 12.5-liter six-cylinder diesel delivered 305 horsepower to the drawbar of its 42,000-pound crawler, useful when pulling a big disc through stubble.

FOLLOWING PAGES: In early August 2014, this 530-horsepower Model 9630 planted grain outside Kalush, Western Ukraine. Deere manufactured these machines in Waterloo from 2007 through 2011.

Dedication

Those who own and operate or collect and show John Deere farm and industrial tractors
know the fundamental truth of Deere's long-time advertising claim: Nothing Runs Like a Deere.
You individuals also know that collectors of other makes of tractor like to suggest
that if you cut a cut, you bleed just as they do. True. But they are colorblind—
this book is dedicated to all those who, when cut, bleed green!

Acknowledgments

The list of individuals who helped create this book is long. Inevitably, I may have omitted someone. Please forgive me if I have and let me know so I can include your name in this section is subsequent printings.

In particular, I am grateful to Harold Brock, the late Chief Engineer, Deere & Co., Moline, Illinois; Lorry Dunning, historian, researcher, and friend, Davis, California; Guy Fay, historian, researcher, and friend, Madison, Wisconsin; Jean Korinke, Development Director, Peter J. Shields Library, University of California, Davis, California; Kevin C. Miller, University Archivist, Interim Head of Special Collections, Peter J. Shields Library, University of California, Davis, California; Rodney Gorme Obien, Archivist + Special Collections Librarian, George C. Gordon Library, Worchester Polytechnic Institute, Worchester, Massachusetts; Jack Schaeffer, Act3 Partners, Tiburon, California; and Les Stegh, retired archivist, Deere & Co, Moline, Illinois.

I especially want to thank the owners, operators, and collectors of the tractors and equipment I have photographed for this book. In alphabetical order, these include:

Bruce Aldo, Westfield, Massachusetts; Steve and Sylvia Bauer, Hastings, Minnesota;

Frank Bettencourt, Vernalis, California; Jim Blomgren, Grand Meadows, Minnesota; John Boehm, Woodland, California; Bettina Chandler, Ojai, California; John Craig, Mentone, Indiana; Paul Cook, Yakima, Washington; Tony Dieter, Vail, Iowa; Sue Duggan, Ostrander, Minnesota; Kenny Duttenhoeffer, El Cajon, California; Todd Erickson, Cannon Falls, Minnesota; Fairfield Equipment, Fairfield, Iowa; Rod Groenewald, Director, Antique Gas & Steam Engine Museum, Vista, California; Junior Heim, Glidden, Iowa; Bruce Henderson, Vail, Iowa; Mike and Pat Holder. Owosso, Michigan; Travis Jorde, Rochester, Minnesota; Wendell and Mary Kelch, Bethel, Ohio; Walter and Lois Keller, Forest Junction, Wisconsin; Bruce and Judy Keller, Brillion, Wisconsin; Mary Jane Keppler, Director, Froehlich Foundation for the Preservation of Tractor History, Froehlich, Iowa; Dax Kimmelshue, Durham, California; Jack Kreeger, Omaha, Nebraska; Harland and Kenny Layher, Grand Island, Nebraska and the collection of the late Lester Layher, Wood River, Nebraska; Brent & Eileen Liebert, Carroll, Iowa; Frank McCune, Newport Beach, California; Don and Charlene Merrihew, Mount Pleasant, Michigan; Cecil Morton, Vista, California; Paul Ostrander and Mike Ostrander, Columbia City, Indiana; the late Robert Pollock, Vail, Iowa; Dave Robinson, Audubon, Iowa; Jim and Fay Slechta, Vail, Iowa; Virginia Schultz, Gary and Charlene Schultz, Ollie, Iowa; Jared and Jeri Schultz, Cedar Rapids, Iowa; Roger and Ruth Swanson, Froehlich Foundation for the Preservation of Tractor History, Froehlich, Iowa; and Roy Volk, Vista, California.

Last but not least, my deep thanks go to Darwin Holmstrom, Senior Acquisitions Editor; Cindy Laun, Art Director; Zack Miller, Publisher; and Jordan Wiklund, Project Manager, at Motorbooks, an imprint of the Quarto Publishing Group, Minneapolis, Minnesota.

Randy Leffingwell
Santa Barbara, CA

Index